Sources for the Study of Science, Technology and Everyday Life 1870–1950 Volume 2: A Secondary Reader

This reader is one part of an Open University integrated teaching system and the selection is therefore related to other material available to students. It is designed to evoke the critical understanding of students. Opinions expressed in it are not necessarily those of the course team or of the University.

Sources for the Study of Science, Technology and Everyday Life 1870–1950

Volume 2: A Secondary Reader

edited by
Colin Chant
at the Open University

Hodder & Stoughton
A MEMBER OF THE HODDER HEADLINE GROUP
in association with the Open University

British Library Cataloguing in Publication Data

Sources for the study of science, technology
and everyday life 1870-1950
1. Science, ca 1850-1976 2. Technology,
to ca 1950
I. Roberts, Gerrylynn K. II. Chant, Colin
III. Open University
509',034

ISBN 0-340-49001-2

First published in Great Britain 1988

Impression number 14 13 12 11 10 9 8 7 6 5
Year 1998 1997 1996 1995 1994

Typeset by Gecko Ltd, Bicester, Oxon
Printed in Great Britain for Hodder & Stoughton Educational, a division of Hodder Headline Plc, 338 Euston Road, London NW1 3BH by Athenaeum Press Ltd, Newcastle upon Tyne.

Contents

Preface

This volume and its companion[1] were compiled for an Open University second-level undergraduate course *Science, Technology and Everyday Life 1870-1950*. The course deals with the effects in everyday life of the complex of technological and industrial changes usefully encapsulated as a 'second industrial revolution': new technology in the shape of materials, power sources, transport and communications media; the institutionalization of scientific research and its technological applications; the spread of mass-production methods and of scientific management, and the emergence of the multi-unit business corporation. Amongst the methodological issues considered are the causal relations of social and technological change, and the ideological content of notions of scientific and technological 'progress'.

In the primary sources collection, it was more appropriate to devote editorial words to the provision of supporting biographical information, rather than try to pre-empt the considerable task of relating a very diverse selection to the main course themes. In the case of the secondary sources, authors have in many cases openly addressed these themes, and so for that and other reasons, it looked more practicable to devise an independent structure, and provide brief commentaries, for the benefit of the general reader. In almost all cases, some editing of book chapters and articles has proved necessary. This has been in part to keep them within the temporal limits of the associated course, and also within its particular focus on science and technology. Some of the extracts were particularly richly endowed in their original form with references to the historical archives on which research had been based. For the pedagogic purpose of this collection, these have been reduced to a minimum consistent with scholarly referencing of sources; or sometimes eliminated, as in the case of Heather Hogan's references to Russian sources, where their value for the general reader would be severely limited.

It should be acknowledged that a production of this sort is very much a cooperative effort, in which academic colleagues on the course production team contribute to the selection and editing of sources, and secretarial and course support colleagues ensure that intentions become actions.

Note

1 Roberts, Gerrylynn K. (ed.) (1988) *Sources for the Study of Science, Technology and Everyday Life: Volume One, A Primary Reader*, London: Hodder and Stoughton.

Acknowledgments

The editors and publishers would like to thank the following for permission to reproduce material in this volume:

Basil Blackwell for 'Technology and Science as "Ideology"' by Jürgen Habermas from *Toward a Rational Society: Student Protest, Science and Politics*, translated by Jeremy J. Shapiro; Cambridge University Press for 'The Movement for Housing Reform in Germany and France' by Nicholas Bullock and James Read from *The Movement for Housing Reform in Germany and France 1840–1914*; Carnegie Mellon University for 'The Revolution in Rural Telephony 1900–1920' by Claude S. Fischer, published in the *Journal of Social History,* Vol. 21, 1987; Associated Book Publishers UK Ltd for 'Childbirth' by F. B. Smith from *The People's Health 1830–1910*, published by Croom Helm, 1979; and 'The "Consumer Revolution" and the Growth of Factory Foods: Changing Patterns of Bread and Cereal-Eating in Britain in the Twentieth Century' by E. J. T. Collins from *The Making of the Modern British Diet* by Derek Miller and Derek Oddy (eds), published by Croom Helm, 1976; J. M. Dent & Sons Ltd for 'The Canker of Industrial Diseases' by Anthony S. Wohl from *Endangered Lives: Public Health in Victorian Britain*; Gerald Duckworth & Co. Ltd for 'The Nature and Uses of Detergents' by W. J. Corlett from *The Economic Development of Detergents*, reproduced by permission of Duckworth; Free Association Books for 'The Reuters Factor' by Michael Chanan from 'The Reuters Factor. Myths and Realities of Communicology: a Scenario' in *Radical Science 16. Making Waves: the Politics of Communications*; and 'More Work for Mother; Technology and Housework in the USA' by Ruth Schwartz Cowan from *Science, Technology and the Labour Process: Marxist Studies*, Vol. 2 by Les Levidow and Bob Young (eds), 1985; Heather Hogan/the editors of *The Russian Review* for 'Industrial Rationalization and the Roots of Labour Militance in the St Petersburg Metalworking Industry 1901–1914' published in *The Russian Review*, Vol. 42, 1983, pp. 163–90; James M. Laux and L. J. Andrew Villalon for 'Steaming Through New England with Locomobile', first published in the Factory, by Reyner Banham from *A Concrete Atlantis: US Industrial Building and European Modern Architecture 1900–1925,* 1986; Oxford University Press for 'The Pokeable Companionable Fire' from *The Politics of Clean Air* by Eric Ashby and Mary Anderson, reprinted by permission of Oxford University Press; The University of Wisconsin Press for the extract from *Frederick W. Taylor and the Rise of Scientific Management* by Daniel Nelson, 112 University of Wisconsin Press, 1980.

1
Materials

1.0 Introduction

Identifying historical epochs with new materials is standard, and justifiable, practice for pre-history, but more contentious when we come to the complexities of modern industrial society. For some, however, the precedent has been irresistible, and even though ancient millennia shrink into mere decades, 'ages' of steel, concrete and plastics have been successively trumpeted. Such metaphors have a clear role to play in any ideology of technological progress, and are (relatedly) highly suggestive of the most simplistic technological determinism. Any determinism of materials (the most vulgar of materialisms?) attracts objections levelled at any technological determinism, notably that materials not only shape society, but are shaped by it in return. In any case, a determinism of materials looks inadequate even as a technological determinism, it being unacceptable to reduce any complex innovation, such as radio, to its component materials alone.

Those who see technology as the historical prime mover are generally blind to other necessary, but non-technological, conditions of change. But it is both undeniable, and historically enlightening, that much social change has depended upon technological innovations. Cinema-going, high-rise living and many aspects of modern continuous flow factory production, for example, are surely inconceivable without celluloid, structural steel and reinforced concrete, to speak only of materials. Some evidence for these propositions is provided in the following extracts from Dennis Sharp and Reyner Banham, on the implications of structural steel and reinforced concrete for the design of cinemas, and factories, respectively. But Banham especially makes it abundantly clear that factory design developed within a dynamic technological and economic framework: the 'daylight factory' is partly, but not wholly, the outcome of a 'revolution in material.'

The extracts on products of the chemical industry also raise questions about the causal relations of technological innovation and social change. The decline in hard soap production in the first half of the twentieth century, described by W. J. Corlett, could be said to reflect, rather than determine, social change in the home and the factory: the rise of soft soap is perhaps a hard case for the technological determinist. Both extracts on chemicals raise more overtly issues about technology and values, and work not only as secondary sources for the period of the Second Industrial Revolution, but also to some extent as primary sources for contemporary evaluations of chemical products. Thus Sylvia Katz, for instance, acts as both historian of public taste, and publicist of plastics and their 'god-like' progenitors. Perhaps the most striking lesson from her piece comes from the contradiction between the consumers' image of plastics in the second quarter of the twentieth century, and the producers' vision of a 'wonder age of plastics' – no clearer indication, surely, of the ideological content of faith in scientific and technological progress, as well as of the untenability of any simplistic determinism of materials.

1.1 Dennis Sharp, *The Picture Palace*, 1969

Drury and Gomersall were the architects for what was described by the promoters as the 'Super Cinema in the Suburbs'. In 1931 the Regal, Altrincham – designed to be 'A CATHEDRAL OF CINEMAS' (shades of Roxy here) – was planted in Manchester's wealthiest and healthiest suburb, with 'a commanding and delightful Terra Cotta frontage . . . majestic and imposing . . . that always retains its "seaside" appearance'. Seaside? Another virtue appears to have been its cleanliness, and the blurb continued 'it can be washed down'.

From the planning point of view this cinema had a number of original features, and of course, as with all the work from the Drury and Gomersall office, it was extremely well constructed and logically planned. The whole building was steel framed – over 200 tons of steel were used – the roof itself being supported by 90-foot long principal girders and the balcony supported by one large girder, 90 feet long and 14 feet deep and weighing 50 tons. The total cost of the building was in the region of £27,000, about the average price for a suburban cinema at the time.

The general colour scheme within the auditorium followed the 'luxurious' pattern, and silver and gold predominated. The inside of the dome in the ceiling of the auditorium gave the impression of being decorated in beaten silver even though in reality it was carried out in plastic paint treated with a metallic finish. The art of the interior decorator was matched by the ingenuity of the electrical engineer in this as in many other cinema buildings, and the use of concealed coloured lighting heightened the total effect.

In this super cinema nothing was spared. For the discriminating audience that the management hoped to entice to its shows, films were booked well in advance, advertised, and programmes were changed every week ('in order that the best of the available "Super Productions" could be shown'). An orchestra was provided (because 'music built up the cinema business'), British Thomson-Houston talking equipment was installed (because it was British), a Compton organ was included (for the same reason) and a large café (in the British rather than the French style) was also provided. The comfort of the patron was also taken into careful consideration: draught screens were erected around all entrances, a confectionery kiosk was provided inside the waiting-lounge, five exits were provided from the balcony and circle, and a car park was situated behind the building. [. . .]

Three broad categories of auditorium arrangement emerged in the 'thirties: the single flat or slightly raked floor type for small cinema buildings, the stadium type with a raised tier at the rear of the auditorium, and the larger 'super' type with a single balcony. There were no similar categories of plan shape and consequently whether a cinema auditorium was fan-shaped, straight back and sides, curved or pear-shaped in plan was dependent on site factors as well as the preferences of individual designers.

Source: Dennis Sharp, *The Picture Palace and Other Buildings for the Movies*, London: Hugh Evelyn, 1969, pp. 95, 125–6, 162, 164–6.

The Regal Super Cinema, Altrincham, Cheshire. Architects: Drury and Gomersall

If the site conditions demanded a long auditorium with a restricted width (as was the case with the Ritz, Hereford, by Leslie Kemp) or a narrow frontage (Regal, Margate, by Robert Cromie) then this very largely dictated the final plan shape. In the Ritz, Hereford, the clever planning of the pre-auditorium spaces enabled the architect to adopt a stadium plan without sacrificing any floor area by tucking the entrance foyer under the steep stalls. In Cromie's Regal the narrow frontage and deep side walls of adjoining buildings were turned to good use, with the whole of the ground floor area given over to a long entrance hall. Above this a restaurant was lit by a glass and concrete laylight. In this example the architect overcame the cramped conditions of the frontage by creating a vertical emphasis with a tower-like façade. Once through the narrow entrance hall the patron came to the spacious foyer and then to the auditorium, which was the usual size for a 1,800-seat cinema. Difficulties encountered in planning inevitably led to difficulties in construction and again it was often the ingenious rather than the conventional structural solution that succeeded.

Modern steelwork design was used almost exclusively in the construction of cinemas in this country. Great balconies could be carried without additional supporting columns obscuring the patrons' view in the back stalls, and immense loads could be transferred to the wall stanchions using a minimum of material. As the important feature of cinema building from the financial point of view was 'the maximum use of available space with the minimum of expenditure on construction' steelwork was ideal.

Reinforced concrete construction found little favour. It could not provide the advantage of steel in either speed of erection, flexibility or dryness. Although a patent was taken out for a system of dry pre-cast concrete construction which would have dispensed with the need to plaster internal surfaces, this did not catch on in cinema building although it was used successfully in flat construction at the time. The advantages offered by steel-framed and 'composite' construction (brick bearing walls and steel trusses) rapidly became obvious. It was economic, relatively easy to handle and fix, and it could be easily boxed in and disguised with fibrous plaster. In being able to span reasonable distances it offered an easy solution to the problem of creating stage areas and proscenium openings. The largest of the supers could be erected within twelve months.

The Kensington Cinema by Leathart and Granger was one of the earliest post-war steel-framed buildings in which the architects dissociated the main auditorium structure from that of the balcony. This gave the immediate advantage of being able to roof-in the building and build up the balcony under satisfactory working conditions. The main walls and roof were constructed from arches resting on pin-bearings at foundation level with wall and roof panels filling the spaces between the arched supports.

In the huge Dreamland Cinema at Margate, designed by Iles, Leathart and Granger in 1935, an even more adventurous use was made of steelwork. Here it was necessary to bridge the auditorium across a main approach road to the amusement park. The main balcony girder (which weighed 60 tons) spanned almost across the axis of the road.

The block-like appearance of many English cinemas was a direct result of the use of structural steel. Widths and heights of structural bays dictated the final form of buildings by the division of the plan into

The Kensington, London, 1926. The auditorium.

The steelwork of The Kensington

The Dreamland Cinema, Margate, Architect: Julian Leathart.

three main elements – entrance block, auditorium and stage. Another factor that seriously affected the external appearance of cinema buildings was the choice of a suitable shape for the roof line. The flat roof was the first choice of most architects both for aesthetic reasons and because it was easy and cheap to erect, particularly after the introduction of lightweight steel decking. It was also found possible with the flat roof to step-down the roof where the auditorium ceiling sloped towards the proscenium and to reduce the volume – and therefore the total cost – of the building. This was done by Leathart and Granger in the Sheen Cinema. The pitched roof however had a currency in entertainment buildings and when finished with asbestos sheeting provided the cheapest form of cover available. More often than not it was used by designers with little aesthetic sensibility or on a building that presented only its main façade to a street. On an island or corner site its use was always visually disastrous.

To a very large extent all the foregoing factors controlled the total shape of individual buildings. What was completely left to the imagination of the architect was the choice of facing materials, lighting, decoration and detail planning.

With composite construction – and with the English climate in mind – most of the early cinemas were in facing brickwork or reconstructed stone. Later, as new materials became readily available and designers began to realise that they were not necessarily building for posterity, less permanent finishes came into vogue. Colour was introduced into large areas of the façades, and biscuit-coloured faience tiling, terrazzo, terra-cotta used as a veneer, white pre-cast concrete slabs and black 'vitrolite' were among the new materials introduced.

1.2 Reyner Banham, *The Daylight Factory*, 1986

It seems clear that the emergence of the flat-topped profile was a mostly American affair and came in with the general rationalization of factory and mill building procedures in the 1820s that, together with the use of a heavy timber internal frame for fireproofing reasons, can be lumped together as 'regular mill construction.' That term applies, technically, only to the timber internal frame, but it can stand here for a significantly American development in the construction of large industrial works.

Before that time, American industrial building was little differentiated from the general tradition of large-scale utilitarian constructions that runs back through the mill buildings of early industrialization in Britain and beyond that to the brick and stone warehouses of the Baltic and North Sea ports. Described in words, it is liable to sound like nothing more than the way people had to build anyhow; one could properly say that, in England, a Georgian house and a Georgian factory would, of necessity, be built in the same way. And not just in England; the first premises occupied in 1875 by the Larkin Company, on Chicago Street in Buffalo, illustrated in Siegfried Giedion's *Space, Time and Architecture*[1] and still standing during my time in that city, was built of brick and

Source: Reyner Banham, *A Concrete Atlantis: US Industrial Building and European Modern Architecture 1900–1925*, Cambridge, Mass.: MIT Press, 1986, pp. 39–44, 56–65.

timber with segmented door and window openings, just like a regular eighteenth-century row house in Boston or even London.

Nevertheless, one can sense the emergence of a separate and parallel tradition, even in Europe, for the larger type of utilitarian structure well before the end of the eighteenth century; J.M. Richards's classic study *The Functional Tradition* (1958) abounds in examples that could never be mistaken for domestic structures even though the constructional procedures and all the architectural elements are identical. The evolution of the alternative tradition may well have its origins in dockside warehouses. The oldest I have seen are on the Nyhavn in Copenhagen, Denmark; the earliest were built before 1700. They are of brick, often with protective stone trim on the ground floors; their windows have regular segmented heads; the internal structures are of heavy timbers with, however, comparatively narrow bay widths of ten to twelve feet.

Their most notably 'industrial' aspect, apart from their lack of ornamentation, is their height, which may be as much as six stories, under a tall and very visible tiled roof, which is their most obvious difference from their later American descendants. Their great height must have arisen largely from two obvious economic considerations: competitive pressure on expensive land at the water's edge and a relatively primitive technology for moving goods. The two things work together; what made waterside land so valuable was the comparative diseconomy of moving goods any distance horizontally on land; the traditional ratio that I was taught in grammar-school geography is that it was ten times more costly to move goods by land than by water, so there would be good reasons for locating the warehouse as close to the ship as possible.

In addition, moving goods vertically by cranes was a good deal more convenient, and capital intensive rather than labor intensive, than moving them horizontally in carts or on pack animals. This was certainly one of the main recommendations of this building format as a way of housing manufacturing industry. It also produced the most conspicuous variation from a perfectly regular system of identical fenestration from end to end of the building: the insertion of vertical stacks of larger openings to serve as hoisting doors through which goods could be swung in or out when hanging from the crane, which was usually a simple, swivelling, triangular bracket mounted above the highest door.

That kind of purely verbal description could apply almost equally well, two hundred years later, to structures such as housed the Buffalo Weaving and Belt Company plant of 1896. However, the physical structure of the version that appeared there alongside the Belt Line Railroad tracks, which made the suburbanization of Buffalo industries possible, had been transformed by vast refinements, subtle improvements, and, above all, rigorous rationalization. For, looking along the more than quarter-mile sequence of contiguous buildings occupied by Buffalo Weaving and built over a period of less than fifteen years, one could see exactly what rigorous rationalization could do to industrial building in an era of improved horizontal transportation. Ultimately, it would kill off the multi-story type with its cranes and hoisting doors (which survived vestigially on the tracks side of the oldest block) and replace it with the single-story workshed, where everything moved horizontally from truck dock to truck dock.

While such improved transportation was lacking, however, or until its potential was properly understood, the multi-story factory reigned supreme. It gave the world its standard imagery of the horrors of 'Smoketown' with its 'dark, satanic mills,' but it also gave the standard images of prosperity: the panorama of row upon superimposed row of regular lighted windows, under the smoke belching merrily from hundreds of smokestacks ('merrily' because it meant all boilers were fired and everyone was working). Throughout northern England, Lowland Scotland, large tracts of northwestern Europe and almost the whole of the north-eastern United States, the multi-story mill or factory flourished as one of the most successful (in terms of Darwinian survival) vernacular building types in the recent history of architecture. Local variations and detailing notwithstanding, the demand for rational construction and rationalized production processes, combined with the need for compact plans, meant that whether built of brick or stone, with an internal structure of wood or iron, its overall form would be pretty well invariant, wherever it stood upon the Earth's surface.

The compact type of plan, inherited, along with its multi-story height, from its warehouse origins, was of peculiar and increasing value to industry as mechanization and concentration increased. During the period of water mills as prime movers, the efficient use of the power from the mill wheel required that the works be built close to the mill race and dam. This could sometimes put as severe a pressure on difficult sites as had been the case with dockside building. Ideally the plant should be built right over the dam, and competition for such locations, in narrow and precipitous mill valleys, was so sharp that in parts of New England, for instance, usable mill sites became known as 'privileges' and their use was formally legislated.

Two other considerations had greater consequence: distribution of power within the plant and the availability of light at the individual work station. While power from the mill wheel, or later the main steam engine, had to be distributed by shafting throughout the works and then taken off by belts and pulley wheels to individual machines, power losses through friction, etc., could best be reduced by keeping the number of shafts to a minimum, ideally one main power shaft per machine floor. This, inevitably, encouraged long buildings of comparatively thin section with the machines disposed along the power shafts, frequently in a double file on either side of them. The need for adequate lighting also encouraged fairly thin buildings, since this would put more of the work stations near an outer wall and thus, with luck, near a window.

Even when buildings were very large, as in the highly rationalized mills in Lowell, Massachusetts, they tended to be only four or five window bays thick. Such depths were of little more than domestic scale, as were the story-heights (twelve to fifteen feet); and the total heights of such structures – four to seven stories was the common range – were again little beyond the domestic range as exemplified in tenements or apartment blocks for urban locations. The workplace was not, in many ways, all that different from the buildings where the workers ate and slept and raised their families. Those who know Raymond Boulevard, along the river in Newark, New Jersey, will recall a sequence of three multi-story brick buildings, all remarkably similar in build and appearance: two of them are factories, but the other, number 627, is a well-made tenement for workers.

Buildings of this type, the 'regular mills' of 'regular mill construction,' were the 'former tradition' to which reference was made at the beginning of this chapter. Because they were a sound economic proposition, they existed long enough as individual structures – and as a type – to become a tradition. Available construction methods and materials had already been pushed almost to their limits when the basic type was stabilized in the 1820s, so it was unlikely to be disturbed by any major technical developments until new materials became available. Its long life was a tribute to its fundamental suitability and adaptability, but its limitations were far more severe than may be apparent to the modern visitor, who tends to encounter it only in its glamorized and romanticized afterlife, gussied up as a stylish boutiquerie, or with its ancient bricks scrubbed naked and its noble timbers educationally exposed in their last mortal incarnation as the Visitor Orientation Center of an Historic Industrial District.

The brightness of modern electric lighting effectively, paradoxically, and appropriately obscures what was clearly perceived as the main defect of the regular mill type in the days when less efficient forms of illumination were all that could be had: inadequate daylighting. Even before regular mill construction was finally replaced by concrete-framed structures, the search for wider windows forced modifications of the external walls of the regular mill that drove the structural performance of brick and masonry to their last limits. In the process, the wall underwent a kind of terminal transfiguration which, almost coincidentally, prefigures the concrete frame and is indeed imitated in some transitional examples of concrete construction.

In this final version, the solid wall pierced by windows (the kind of wall that Ransome had effectively imitated in concrete) was replaced by a system of separate brick columns, connected by thinner membranes of brickwork containing windows that went almost from pier to pier. In this way the weight of brickwork that had to be supported by the arch or beam that spanned each opening could be reduced to the minimum required for decent weatherproofing, but the stability of the wall as a whole was guaranteed by the thickness of the piers. [. . .]

Clean, well-lighted places

However secure the former tradition, with its habits of adaptive pragmatism and its inherited vocabulary of architectural forms, may have appeared at the very beginning of the present century, it was to be overthrown by a genuine revolution in material and concepts between 1902 and 1906. The nature of that revolution and the profundity of its consequences are epitomized within the vast bulk of 701 Seneca. A couple of years after the completion of the main building campaign in 1907, the company cleared the site of the last remnant of the previous generation of buildings, the old boiler house, which had been replaced by a new and large power house immediately to the west of 701. The cleared site was little more than a notch in the back wall of the monster, seven stories high but less than fifty feet wide, and the company filled it with the new Building C to render the work floors continuous across what had been a profitless interruption.

Against the grain of all the previous brick-pier construction by the Reidpaths on the 701 Seneca site, the company's own design staff, and

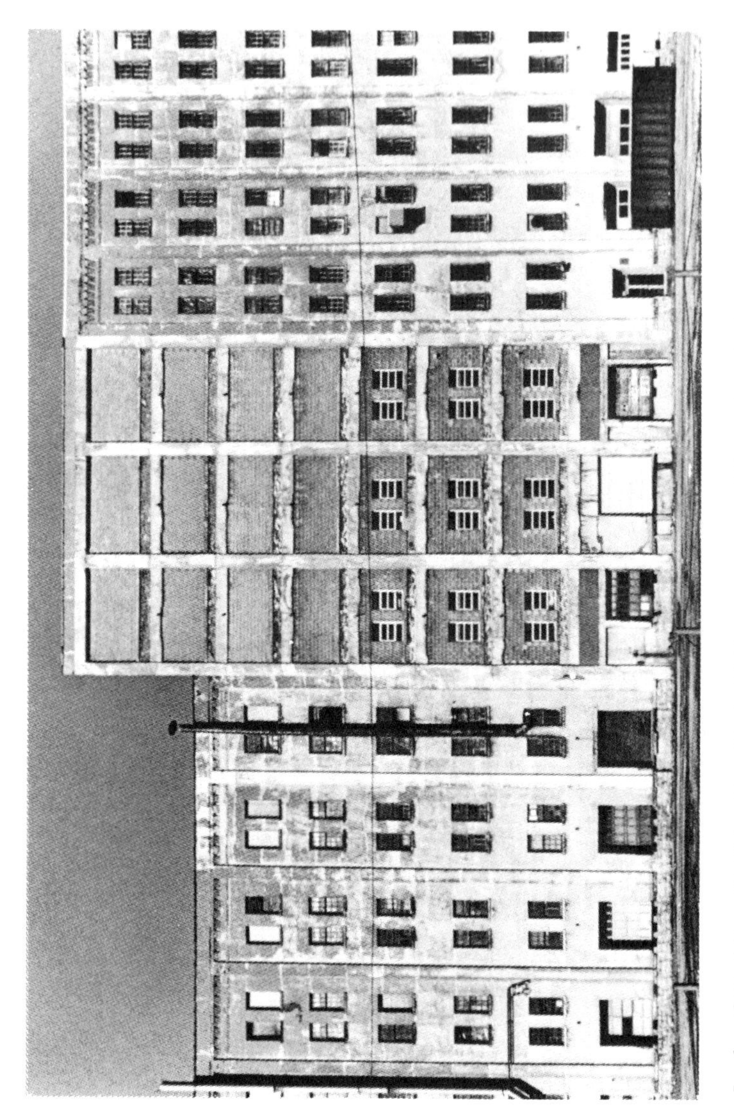

Larkin C Building, Buffalo, New York, 1913, present state. (Photo, Bazelon)

Larkin N *Building, Buffalo, New York; interior showing daylighting through brick exterior wall. (Photo, author)*

Larkin C *Building; interior, showing level of daylighting using reinforced concrete frame. Photograph taken at the same aperture and shutter speed as the picture above and under identical external lighting conditions.*

their contractors, devised three bays of concrete-framed construction of absolute simplicity. The openings to the outside were glazed clear across from column to column and from the underside of the floor slab above down to a sill that topped a low brick spandrel wall just high enough to carry a hot water radiator. The environmental transformation wrought by this innovation could be seen at 701 Seneca as nowhere else in the world, because the fully glazed concrete insert of Building C was entirely surrounded by works in the former tradition – with windows tunnelled through a massive brick wall – and one could walk through directly from the old into the new or vice versa. The difference in daylighting levels could only be described as sensational. Whereas natural light of usable strength barely penetrated the depth of one fourteen-foot bay from the twinned windows of the brick-pier section, it penetrated better than twice that depth from the wall-to-wall and ceiling-high glazing in Building C, and with far less glare because there were no immediate contrasts of juxtaposed light and dark to distress the eyes. [. . .]

By 1911, the year in which the Larkin Company decided to make their last section of 701 Seneca out of concrete, the virtues of the Daylight Factory were just beginning to appear as received wisdom in professional and trade magazines:

> Pure, clear, uncolored daylight – the sunshine of roofless fields which doubtless contributes in no small measure to the rare artisanship of Japanese, Indian and Italian handiwork – is becoming the possession of the American factory laborer. Steel sash windows which are weatherproof, give access to a maximum of daylighting

wrote *The American Architect and Building News* in a mildly poetic vein not unknown in its editorials of the period. It went on to assert that

> If the books were to be examined and an accountant rendered a verdict, the issue might not be entirely clear or decisive, but if the votes of the people connected with the factory were taken there would be no question . . . but that whatever additional expenditure may have been required was fully justified in the improved health, the improved moral, physical and aesthetic conditions.[2]

Though the revolution in industrial building went well beyond daylighting as such, in many eyes that was clearly the outstanding innovation. To provide daylighting of the kind and volume that was now expected, a steel-framed building would need a steel-framed exterior as well, and that, for fireproofing as well as less tangible reasons, seemed to be virtually inconceivable in the first decade in the twentieth century. Yet it had been done here and there: as early as 1898 the old Veeder cyclometer plant at Hartford, Connecticut, had a pure steel frame lightly clad in brick. Furthermore, 'in the design of the building, one of the principal ideas was that it should have an abundance of light on every square foot of floor space, so that it could all be available for carrying out the very fine operations required for the manufacture of cyclometers.'[3]

This must have been one of the earliest successful attempts to build a true Daylight factory, one in which every square inch of the exterior envelope that could reasonably be glazed was filled with glass. [. . .]

Speed and economy were put first in the fairly impressive roster of improvements that were supposed to accompany the change to concrete-framed construction, in the view of the editors of *The American Architect and Building News*. They listed the advantages of the new materials as 'first, speedy erection; second, low first cost; third, daylight illumination; fourth, [earthquake] shock-proofness; fifth, maintenance economy; and, sixth, fireproofness.'[4] For preventing earthquake damage, concrete offered few decisive advantages over properly designed steelwork, to judge from what survived, say, the massive shocks of the San Francisco earthquake of 1906. A case could still be made for steel construction in Daylight factories, as the Veeder plant shows. And the question of maintenance remains somewhat moot, given the short first-user life that most industrial construction in the United States in the twentieth century has usually enjoyed. Nearly all such buildings have normally been neglected from the time their original owners decided to dump them on the market, and the results reveal all too clearly that in most industrial environments, exposed concrete needs far more upkeep and protection against destructive atmospheres than its early champions were usually prepared to admit.

This, then, really leaves only fireproofing as the critical consideration in determining if the desired Daylight factory would be built of concrete. Experience showed that metallic construction usually had a far worse fire performance than the heavy timbers of regular mill construction. Timber deserved its epithet of slow-burning construction and retained its structural integrity for much of the time it was slowly burning, whereas noncombustible metals proved to be dangerous under conditions of great heat: cast iron would shatter, wrought iron and steel would distort and deform. Of course steel could be fireproofed, otherwise regular skyscraper construction would have been far too dangerous to be acceptable or – what was more vital – to be insurable. But the cost of fireproofing skyscraper construction was largely concealed by the fact that social and architectural custom required such structures to be clad in ornamentation without and have tolerable surfaces within. It was only when Mies van der Rohe wished to exhibit the steel frames of his Chicago towers in the 1950s that the fireproofing of skyscrapers became an issue rather than a by-product of customary practice.

In the industrial buildings of the former tradition, such amenities were minimal on the exteriors and nonexistent for interiors, so the cost of wrapping structural steel in gypsum plaster inside, and brick, terra-cotta, tiling, or patent cement outside, was a visible budget item that might be tolerated for public facades or street fronts but nowhere else on the structure. The concrete frame avoided the expense of such added protection by being inherently fireproof, and probably ingratiated itself with many industrial builders for another and apparently unconnected reason: the dimensions for a reasonably strong, economical, and fireproof concrete upright usually came somewhere near to the fourteen-inch square dimension that everybody already knew well from slow-burning timber construction and from fireproofed steel-work. When all these considerations of speed, economy, daylighting, maintenance, seismic stability, and so forth are added to fireproofing, there was a small advantage for reinforced concrete construction in industrial building, an advantage that would probably have grown at some cautious pace in step with industry's growing familiarity with the material and confidence in its reliability but for a dramatic event that seems to have changed opinions coast to coast.

In 1902 the East Coast plant of Pacific Coast Borax, at Bayonne, New Jersey, went up in a spectacular fire that attracted national attention because it was so hot that steel twisted and iron melted into shapeless puddles on the floor. But the floors survived and so did the internal columns and external walls because they were all made of the same material, fireproof reinforced concrete. There could have been no more convincing demonstration of the virtues of the material, but over and above the incontrovertible (and photographable) physical fact of the building's almost complete survival, the concrete industry took full advantage of this free gift of favorable publicity. The Atlas Portland Cement Company saw to it that this would be the best-known conflagration of the new century and gave it a full chapter in their 1907 publication *Reinforced Concrete in Factory Construction*. At the back of the volume is a lengthy testimonial letter, which deserves extensive quotation. After conventional salutations appropriate to a solicited testimonial, the manager of the eastern division of Pacific Coast Borax writes:

> Among some of the special features that occur to us are:
> First: Its being absolutely fireproof. This was fully tested as you well know by the fire which we had in our Calcining Department . . . a fire of terrific heat – melting all exposed metal and burning all combustible partitions etc., that the building at that time contained; but the concrete building itself stood the test magnificently, and as our property is surrounded by stills of the Standard Oil Co., this is particularly important to us, and we know that our building is absolutely fireproof.
> Second: Cost of repairs. No expenditure under this heading is made – the building being monolithic and like Spanish Wine, improves with age.
> Third: Strength. As you know we carry terrific loads on our floors – on our fourth floor carrying a weight of 1430 lbs. per sq. ft. On the lower floors we have carried much heavier weights without straining the building in the least.
> Fourth: Cleanliness. Your construction is an ideal construction for a factory as it can be kept perfectly clean – it being a simple matter to hose and wash it out.
> We believe that concrete construction is the proper construction and that the Ransome system is the best system.[5]

Not surprisingly, this testimonial had originally been solicited by the Ransome and Smith Company, by then established in New York.

The Pacific Coast Borax fire was, it appears, the triumph and vindication of Ransome's professional life. That company's building at Bayonne, erected in 1897, had been his first work on the East Coast and is also reputedly the first complete reinforced concrete factory to be erected on that side of the country. The fire and Ransome's great and growing reputation as an inventor and constructor combined to give a kind of charisma to reinforced concrete as the material of the new industrial age; and Ransome was only one of a number of forceful new engineering personalities who appeared upon the scene as exponents and exploiters of this seemingly miraculous material. [. . .]

Notes

1 Giedion, *Space, Time and Architecture*. 5th ed. 1967, p. 357.
2 *American Architect and Building News*, XCIX: 1851 (June 14, 1911): 214.
3 *Engineering News* XL (July 7, 1898): 15.
4 *American Architect and Building News*, p. 213.
5 Atlas Portland Cement Co., *Reinforced Concrete in Factory Construction*, 1907, p. 237.

1.3 W.J. Corlett, *The Nature and Uses of Detergents*, 1958

When we try to compare soap and the synthetic detergents in order to see the advantages and disadvantages of the different types of product as detergents, we find that, because of the number of factors involved, it is not possible to reach many definite conclusions. One reason for this is that the effects of the detergents depend to a very large extent on the nature of the dirt to be removed and the nature of the surface or material to be cleaned. A second reason is that, while such physical properties as the reduction in surface tension, the amount of foaming, and the powers of emulsification and suspension are all relevant to detergency, there does not seem to be any relation between these properties and detergency which is sufficiently close for their measurement to be used as an indication of detergency. To a large extent, the assessment of detergency has rested on the results obtained on washing test pieces of fabric or dirty dishes. There is still the difficulty that the standard methods of soiling articles for laboratory experiments may not reproduce the type of soiling met with in actual use, while the use of articles soiled in the normal way may introduce very wide sources of variation within the experiment.

The main method of action of detergents can be summarized as follows:

The surface or fabric to be cleaned is made wet; the surface-active agent helps here by enabling the liquid to spread and to penetrate more easily because of the lower surface tension and interfacial tensions. The dirt on the surface is loosened, again largely by the effect on the interfacial tensions, and can easily be displaced from the surface by agitation. Oil in the dirt is emulsified, surrounded by molecules of the surface-active agent, with the molecules orientated on account of the existence of the hydrophobic and hydrophilic parts; solids are taken into suspension. With a good detergent the oil will remain emulsified and the solids suspended and they can be rinsed away, even though the concentration of the detergent in the solution is likely to fall considerably during the rinsing process. If the dirt is redeposited on the surface or fabric and does not get removed in rinsing, or if any other precipitate is so deposited, then the detergent will not be performing its task properly.

Source: W. J. Corlett, *The Economic Development of Detergents*, London: Duckworth, 1958, pp. 19–26, 29–31.

The part played by lather in this process seems to be a little uncertain. It probably has some cushioning effect in protecting fabrics from mechanical friction during washing and so may have some effect on the amount of shrinking. Changes in the amount of lather during washing indicate changes in surface tensions and thus in the concentration of the surface-active agent. It is not, however, possible to compare different detergents on the basis of the amount of lather; nor is it possible to say, in general, that the point at which lather disappears is the point at which the detergent solution ceases to operate effectively. Lather may also help in the suspension of the dirt. It can be a nuisance, as, for example, in washing machines, or a positive advantage, as in the cleaning of carpets without removing them from the floor.

We can now consider some of the technical advantages claimed for soap and synthetic detergents. It is generally agreed that with soft water and alkaline conditions, soap is a good detergent. It suffers from disadvantages, however, in hard water or in acid solutions. In the former, the soap reacts with the calcium and magnesium, which cause the hardness in the water, forming calcium and magnesium soaps (the calcium and magnesium salts of the fatty acids). As these are insoluble, they tend to get deposited on the article being washed; a large amount of soap can also be wasted in overcoming the hardness in the water in this way. In acid solutions the soap decomposes and precipitates free fatty acid. With hard water, it is possible either to soften the water before washing or to incorporate builders in the soap, which will overcome the tendency to deposit the calcium and magnesium soaps. When soap has to be used in acid solutions, the difficulties cannot be overcome so easily.

With many synthetic surface-active agents the powers of holding dirt in suspension are not very good and it is necessary to overcome the tendency to redeposit the dirt with builders just as, in the case of soap, builders are useful in overcoming the hardness of the water. [. . .]

With soap solutions, the amount of foam is usually a good indication of the concentration of the soap and the disappearance of the foam is closely connected with the end of the effectiveness of the solution as a detergent. As we have seen, this is not an invariable property of detergents and does not apply in the case of many synthetics. The greater variability in synthetic detergents, with the large number of different products, makes it easier to get a product with very little foam when this is desirable.

This greater variability in foaming is a particular case of one of the main differences between soap and synthetics, which can be both an advantage and a disadvantage for either. On the one hand, soap is more versatile both in its detergent and non-detergent uses than any particular synthetic; on the other hand, the greater diversity of the synthetics and the wide variation in their performance for different molecular structures make it easier to find the product with the right properties for one particular purpose.

Soap has the advantage that it is much easier, in general, to manufacture; much less technical knowledge is required.

Two other points of difference which have been commonly discussed are the effects on the skin of the user and the problems of the disposal of the materials after they have done their job as detergents. Both of these problems have been considered by the Committee on Synthetic Detergents appointed by the Ministry of Housing and Local Government. The Committee decided that there was no cause for alarm on the first count; that, while synthetic detergents can cause dermatitis when used regularly, there is no evidence that they do so to a significantly greater extent than soap products.[1] The majority of the Committee did, however, believe that the effects of the disposal of synthetic detergents through sewage were sufficient to justify concern.[2] The reasons for this concern were that the synthetic detergents commonly used domestically have frequently led to copious foaming at sewage works, that they appear to have had some adverse effect on sewage treatment and so on the general level of the sewage effluent to rivers, that they are not themselves completely removed or oxidised so that they pass into the rivers surface-active agents and phosphates, both of which can have deleterious effects, and that they can persist through the treatment at waterworks and so be present in small quantities in water supplies. Soap does not give rise to these difficulties since it is easily treated by the normal methods of sewage disposal; some other types of synthetic surface-active agents are also amenable to treatment at sewage works – the Committee state, for example, that the 'particular kind of secondary alkyl sulphate' which is the next most commonly used surface-active agent in this country does not appear to persist through the sewage treatment.[3]

Before the 1939-45 war almost all the detergents produced in the United Kingdom were soap products. Some primary alkyl sulphates had been made for several years; powders for household use and shampoos based on the synthetics were produced for a short time but only on a very limited scale. In looking at the statistics of detergent production there is little loss, for this period, in confining attention to soap (see Table I). Several sources are available although, unfortunately, the figures are not always strictly comparable and are not as complete as one might wish. [. . .]

The most striking points brought out by Table I are the very rapid increase in production and consumption, even after allowing for the increase in the population, during the nineteenth century, and the rather slower increase between 1900 and 1938, particularly during the first thirty years of this period. [. . .]

A little can be said to supplement Table II. Most of the production before 1907 was of hard or soft soap but there had been an appreciable production of soap powders for many years. According to R. L. Wilson, the production of soap powder by R. S. Hudson started in 1863,[4] while C. Wilson states that it was a substantial business by 1875,[5] and that their sale in 1899 was 19,000 tons.[6] The census for 1935 gives the production of soap powders separately as 107,000 tons; in 1938, according to an estimate of the Ministry of Agriculture, Fisheries, and Food, it was 132,000 tons. Even if one allows for differences in coverage of the two figures, this suggests a rise of about 20 per cent in the output of soap powders while the production of all soap products was rising by about 7 per cent. The 1938 estimate of toilet soap production at 41,000 tons shows a very similar percentage increase since 1935 to that of soap powders. Most of the production shown in Table II as 'Others' is of abrasives (about half the total in 1924 and 1935) or of soft soap.

The most striking change shown by all these figures is the decline in the production of hard soap after 1924. In that year it was, by weight, about two-thirds of the total; immediately before the 1939-45 war it was

Table I
Production and Consumption of Soap in the UK, 1821–1938

	Production (000 tons)	*Consumption (000 tons)*	*Consumption per head (lb. per annum)*
1821 (*a*)	44	42 (*b*)	4.5
1851	97	87 (*b*)	7.1
1881	215	200 (*b*)	12.8
1900	355	320	17.4
1912	430	365	18.0
1924 (*a*)	490	425	21.2
1930	475	430	21.0
1938	545	520	24.5

(*a*) Figures for the years 1821–1912 are for Great Britain and Ireland; for 1924–38 they are for Great Britain and Northern Ireland.
(*b*) Imports were not recorded separately before 1900: no allowance is made for 1821 and 1851, but a small allowance is made in 1881.
Note on Table. Figures for 1821 and 1851 are taken from excise duty returns; those for 1881 and 1900 from data given by C. Wilson (the 1881 figure being an interpretation of a rough statement); those for 1912, 1924, and 1930 from the Censuses of Production with a rough adjustment where necessary, for firms not covered by the census; the figure for 1938 is obtained from data in the *Monthly Digest of Statistics*. Import and export figures are taken from the *Annual Statement of Trade of the UK*. All figures for production and consumption (except for 1821 and 1851 when the total was small) have been rounded off to the nearest 5,000 tons but cannot claim to be accurate even to this extent.

Table II
Production of Soap Products in the UK, 1907–35 (000 tons)

	Hard Soap	*Soap Flakes and Soap Powders*	*Toilet Soap*	*Others*	*Total*
1907 (*a*)	283	45	14 (*c*)	42 (*d*)	384 (*d*)
1924	325	80	24	57	486
1935 (*b*)	268	149(*e*)	33	49	499

(*a*) Great Britain and Ireland.
(*b*) Excludes small firms.
(*c*) Includes shaving soap in 1907.
(*d*) Includes soap base in 1907; soap base is excluded from the figures in other years.
(*e*) Includes chips.

Note on Table. All the figures are taken from the Censuses of Production with the following exceptions: The production of soap flakes and powders in 1907 is taken from C. Wilson, *The History of Unilever*, Vol. I, p. 120; the figure for others in this year is the balance to make up the total. The figure for soap flakes and powders in 1924 is estimated since they were not shown separately in the census. Since, however, the larger of the categories not shown separately in the table were given in the census, it is unlikely that the error can be large. Once again, the figure for others is the balance.

only about half.[7] Side by side with this change there was the increasing importance of soap powders and soap flakes. In 1924 they were only about a sixth of the total; immediately before the war they were nearing a third. In addition, there was an increase in the production of toilet soaps from about 5 per cent of the total in 1924 to about 7½ per cent before the war. If one compares the figures for 1907 and 1924 in Table II, one can see that the changes after 1924 were a continuation of what had been happening earlier, although they were more rapid. It can be said that soap powders and soap flakes were cutting into the production of hard soap for one use and that toilet soap was similarly cutting in for another. [. . .]

During the first two or three years of the 1939-45 war the production of soap was not much affected; in 1942, however, the output was reduced and rationing was introduced. At about this time, the Shell group started producing the secondary alkyl sulphate, Teepol, and ICI the non-ionic detergent Lissapol N. A number of soap substitutes, some of them containing a synthetic surface-active agent, were sold on the domestic market; statistics of their total sales are not available but it is improbable that they were very large. The reduction in the output of soap did not last for long and by 1945 output was back to the pre-war level.[8] In the following three years, however, when there was a shortage of oils and fats, it was lower than it had been at any time during the war and 15-20 per cent less than immediately before the war. In the last of these years, 1948, the first of the important synthetic detergent powders was put on the domestic market in the UK. It is, then, useful to compare the soap position in that year with pre-war years, even though the immediate effects of the war were not over – rationing continued until September 1950.

Output of soap in 1948 was 470,000 tons as compared with 545,000 tons in 1938; exports were down slightly from 36,000 tons to 28,000 tons and imports from 9,000 tons to negligible amounts. The decrease in production had not, however, affected all types of product equally. The output of soap powders was about the same as before the war, that of toilet soap slightly higher, while that of abrasives, at 59,000 tons, was substantially higher.[9] The output of soap flakes had apparently dropped somewhat but most of the decrease was concentrated on hard soap, the output of which, at 172,000 tons, was now only about 37 per cent of the total as compared with about 50 per cent before the war. The decline in the importance of hard soap, which was apparent from the pre-war figures, had continued over the war and immediate post-war years.

The division of soap between industrial and domestic use (or between the different types of outlet) had not changed substantially, although both sources of information suggest that a slightly higher proportion was going into industrial use. It must be remembered that soap was severely rationed in 1948.

Since 1948 consumption of both soap and synthetic detergent products has been large. As a result, changes in the figures for soap cannot be understood except when considered alongside those for synthetic detergents. Figures for the production and consumption of of soap are given in Table IV.

It can be seen from the table that there was a considerable increase in the production of soap between 1948 and 1950 when rationing ended but that there was a substantial fall in the next few years. This fall had

Table IV
Production and Consumption of Soap in the UK, 1948–55

	Production (000 tons)	*Consumption (000 tons)*	*Consumption per head (lb. per head)*
1948	470	442	20.0
1949	538	495	22.2
1950	616	572	25.5
1951	567	501	22.3
1952	500	437	19.4
1953	469	414	18.3
1954	456	396	17.5
1955	467	406	17.8

Note on Table. Production figures based on those obtained by the Ministry of Agriculture, Fisheries, and Food or, later years, the Board of Trade. In the earlier years the figures more correctly relate to deliveries by soap-makers than to production. Consumption figures allow, once again, simply for imports and exports and not for stock changes.

eased off in the last year or so of the period covered by the table. The peak shown by the figures for 1950 may be somewhat exaggerated, since there may have been some building up of stocks in the shops in preparation for the end of rationing. Stocks may have been increased to a greater extent than was necessary since many people believed that the synthetic detergents would lose their market when soap was not rationed. In addition, of course, consumers may have increased their stocks.

Figures for the consumption of synthetic detergents in the UK are shown in Table V; these are only available from 1949. They are given on 100 per cent active basis; estimates of the actual weights of the synthetic detergent products are not published officially.[10] Most of the unofficial estimates of consumption of synthetic powders and liquids give

Table V
Consumption of Synthetic Detergents in the UK

(000 tons of active material)	*Total Consumption*
1949	13.0
1950	15.2
1951	17.8
1952	26.2
1953	35.0
1954	39.0
1955 (estimated)	40.0

Note on Table. The figures are taken from the *Report of the Committee on Synthetic Detergents* (Ministry of Housing and Local Government), p. 6.

figures which are approximately five times those given in Table V.[11] These higher figures are the ones which are most nearly comparable with those for soap.

Notes

1 *Report of the Committee on Synthetic Detergents*, pp. 8-10.
2 Op. cit., p. 45.
3 Op. cit., p. 28.
4 R.L. Wilson, *Soap through the Ages*, p. 16.
5 C. Wilson, *The History of Unilever*, Vol. I, p. 12.
6 C. Wilson, op. cit., Vol. I, p. 120.
7 It might be more appropriate to group together hard and soft soap for these comparisons since these are the two traditional washing products. However, since the data for soft soap is incomplete, this has not been done. Soft soap is, in any case, comparatively unimportant in the U.K.
8 Commonwealth Economic Committee, *Vegetable Oils and Oilseeds. 1950*, p. 110.
9 Based on estimates made by the Ministry of Agriculture, Fisheries, and Food similar to those already mentioned for 1938.
10 The classification in the Censuses of Production is rather too wide to be useful.
11 In an ordinary domestic product the weight of active material is about 20-25 per cent of the whole.

1.4 Sylvia Katz, *The Age of Plastics*, 1978

Many people, especially journalists and commentators, seem to think that we live in 'The Plastic Age'. In fact, every decade since the modern plastics industry began in the 1920s has been hailed as 'The Plastic Age'. By right that title is far more appropriate to those early years; new materials were being discovered at an increasing speed and the foundations of the industry as we know it were established. In 1925 *Plastics* had been founded in America, the first plastics journal in the world (later named *Plastics and Moulded Products* and now *Modern Plastics*), followed by a stream of journals in Europe. In May 1930 the first symposium on plastics was held at the Grand Central Palace, New York, attended by 608 people. In London in the same year, one hundred men convened to form a plastics club with the aim of increasing an interest in plastics in England. Finding the word 'Club' somewhat unprofessional, they changed it to 'Institute'; a committee of seven was appointed, and the Plastics Institute was established, now the Plastics and Rubber Institute.

Those early days reverberated with the cries of patent-filers and litigants, a sure sign of a healthy competitive industry. However, as *Modern Plastics* commented, progress benefits better when patents are

---

Source: Sylvia Katz, *Plastics: Design and Materials*, London: Cassell & Collier/Macmillan, 1978, pp. 8–12.

licensed so that know-how can be shared rather than their being contested in lengthy disputes.

In the 1940s the attitude to plastics was no less bewitched: 'There grew up a very general idea that we were on the verge of a "Wonder Age of Plastics",' wrote the Council of Industrial Design in their 1948 issue of *New Home*. This view sprang from the war years, when amazing feats were achieved by plastics. The war nurtured the idea of plastics as 'wonder materials' capable of solving all our problems. Three decades later French retailers are again announcing that *'le plastique est roi'*.

However, as Dr Chilver, Vice-Chancellor of Cranfield Institute of Technology, has said, if any one thing characterizes our age, it is communications, not plastics. The Stone and Bronze Ages are so called because most implements were made of those particular materials. The same cannot be said of plastics today, because plastics are but one of the whole range of necessary materials. Plastics are yet to come of age – an event that will happen, according to J. Harry DuBois, in 1983 when the volume of plastics will equal that of all metals. It is estimated that one third of all chemists and chemical engineers in the West are now engaged on the development of plastics. In 1975 the Science Research Council (SRC) announced the establishment of a Polymer Engineering Centre, to coordinate both processing and design in plastics, while at Lewes Prison, in Sussex, sewing mailbags has been replaced by plastics moulding. Hopefully, however, there will never be a 'Plastic Age' in the proper sense of the term.

The asset of plastics is that, unlike precious metals and stones, it is a universal material, available to everyone. Synthetic resin is transformed into articles that can be exclusive and expensive, or cheap rubbish. Plastics can also be superfluously decorative, but at the same time, according to Charles Spencer, can fill a social need: 'Perhaps there is something in plastics, as a kind of democratic material, capable of the most mundane and the most serious uses, identified with the "space" age, which promotes a down-to-earth, social conscience' (*Art and Artists* London, September 1967). But as Alan Glanvill pointed out at the Society of Industrial Artists conference in May 1975, plastics have suffered from the success of fulfilling this social need, for they have thus descended the social scale in the mind of the user. [. . .]

Plastics – 'cheap and nasty'?

'I was born with a plastic spoon in my mouth . . .
Substitute your lies for fact, I see right through your plastic mac.'
('Substitute' by Pete Townshend, 1966)

Derived from the Greek adjective *'plastikos'* meaning mouldable, the Plastic Arts have traditionally signified crafts which employ materials capable of being moulded, such as metal, clay, plaster and glass. 'Plastic' in this classical sense has meant three-dimensional, formed, plasmic; hence the phrase 'God the great plastic Artist (1741)' *(Shorter Oxford Dictionary)*. We also apply the adjective 'plastic' to materials which are very flexible in their finished state.

The word 'plastics' in the plural, can be used as both noun and adjective. The word was officially recognised in 1926 as a collective noun describing a group of materials capable of plastic deformation. Its adjectival use was accepted by the British Standards Institution in 1951. Thus, 'The Plastic Age' should strictly speaking be 'The Plastics Age', but as 'plastic' is now the commonly used form it has been adopted throughout this book. However, the meaning of the word 'plastic' with which we are all most familiar is to denote something fake and synthetic. The problem with plastics, as anyone concerned with the manufacture, design, or the retailing of plastic goods will endorse, is the myth that still exists that they are a cheap substitute for something better.

Because plastics have suffered from the success of widespread proliferation, which has cheapened and debased them, the mistakes that have resulted merely reinforce the poor public image. Even now shoppers still ask: 'Is this real or plastic?' as if plastics could never be real. Semiotics, the general science of signs, points out that the more often a certain word or phrase is used, the more its symbolic meaning is reinforced, until it is finally accepted as a cultural cliché and is absorbed as a myth. This has happened with the words 'cheap plastic', which have been repeated so often that 'cheap' and 'plastic' have become synonymous, although plastics are becoming real materials in their own right for which there is no substitute. [. . .]

Antagonism towards plastics is so established and its potential effect on marketing so heavy that manufacturers of modern paints dare not print the word 'synthetic' on their tins although most paints now are synthetic. Instead they use euphemisms like 'Silthane Silk' or 'Vinyl Satin'.

The origins of the present-day connotations of the notion of cheapness can be seen in the history of plastics. Crisis points in history, when the world's supplies of natural materials have been threatened by wars, population explosions or economic change, have often been a spur to technological development. The origin of our modern plastics industry stems from such a crisis in the 1840s when the population increase strained the supplies of natural materials such as silk, rubber and ivory. Those materials were already becoming increasingly expensive. Celluloid, the first man-made plastic, was originally patented in America in 1869, as a cheaper substitute for ivory billiard balls. Casein, which was patented in 1897 in Germany under the trade name 'Galalith', developed originally as a cheaper, substitute material for horn and silk. Phenolic resin was often referred to as 'imitation amber'.

World War I created renewed demands on all natural resources, over which governments assumed control, and speeded up the development of synthetics. Before 1914 Germany had pioneered synthetic rubber, and by the outbreak of World War II had prosperous PVC and polystyrene industries. World War II stimulated research in all fields of polymer chemistry. PVC, polythene and 'Perspex' were the only familiar thermoplastics during the war, but after it new plastics began to appear commercially at an increasing rate: polystyrene, GRP, polyurethane, polypropylene, polytetrafluoroethylene and the polycarbonates. The 'miracle materials of war', formerly guarded military secrets, became free to benefit society. Factories that had manufactured war materials returned to their original products and applied increased knowledge of materials and processes.

For example, a factory previously laminating aircraft parts developed new shapes for plywood chair shells using the newly acquired bonding techniques and synthetic glues. Charles Eames' GRP chairs reflect the fact that he once worked on experimental gliders as part of the war effort.

The Britain Can Make It exhibition organized in 1946 by the Council of Industrial Design at the Victoria and Albert Museum, London, exhibited post-war designs alongside war-time precursors, illustrating positive developments derived from the war experience. But this well-applied war experience was counterbalanced by an over-zealous misapplication of plastics. Widespread application during post-war austerity reinforced the disgraced reputation plastics already had and which the industry is still in the process of setting right.

But the war had already damaged the reputation of plastics in other ways. Developments in Germany in the thirties aroused tremendous jealousy, particularly in Britain. Awareness of the certainty of war had pushed forward German research, and with Hitler encouraging *Heimstoffe* (home-made materials), many substitute materials were developed. For example, in 1935 the Germans were using a synthetic phenolic resin, 'Neoresit', to replace the costly amber resin used for making blood transfusion vessels. They created a product which not only did what was required but possessed the added property of being able to be repeatedly sterilized. During the war the Germans used a vinyl solution, 'Persiton', polyvinyl pyrrolidone, as a substitute for blood plasma. It was obviously very successful as its use later spread throughout Europe and it has been stored for national emergencies in America.

At first the British reaction was to jeer. Many of the German experiments were even considered uneconomic. Allied propaganda labelled their inventions as 'synthetic' whereas in fact they were well-researched substitutes, and it is significant that a German word 'ersatz' was adopted to refer to substitute materials. The trouble is that plastics have nearly always been used as a substitute material, and a high failure rate has contributed to plastics' poor image from the early days.

In the thirties plastics were often used in a manner which the purists amongst us would call 'dishonest'. The flood of replica wooden clocks made of 'Bakelite' conceded nothing to the fact that these cases were totally unnecessary: modern clock movements by that time were mostly constructed with tight dust-proof covers. Concern for the response of materials to this treatment was expressed as early as 1933 by Joseph Sinel who was working at that time for RCA-Victor, Westinghouse and Texaco. He complained that the essential quality of 'Cellophane', its transparency, was often totally obscured by copious over-printing, and to counteract this 'common abuse of Cellophane' he advocated 'rational machine aesthetics'.

The golden era of plastics in the early thirties was followed by a period of crisis and concern at the increasing prejudice against plastics due to the sale of inferior products. Bargain prices bought shoddy goods – faded colours, crazed mugs, cracked boxes and designs that came off in the wash. The whole industry became demeaned and the effects were felt all down the line from designer through producer to consumer. [. . .]

Caution, even defensiveness, set in in reaction to the poor image of plastics which became common after their usefulness in the war had been forgotten. Post-war failures led to the British Standards Institution laying down standards for plastics, as the Germans and Americans had done years earlier. The setting-up of the Design Centre by the Council for Industrial Design in December 1944 was aimed at promoting design as a vital part of industry. The Council of Industrial Design slide library has a special category in its index for misapplied plastics, labelled 'Plastics, cautionary', and the lessons are being learned, although mistakes still happen. [. . .]

Perhaps the most fascinating aspect of plastics is their mystery and anonymity, that the residue of billions of prehistoric organisms can be transformed, through a series of complex processes, into an endless variety of products. 'Industrial chemistry today rivals alchemy. Base materials are transmuted into marvels of new beauty,' wrote Paul T. Franklin in 1931.

Today these synthetic resins no longer try to simulate precious materials as they once did. Plastics are used so as to declare openly their innate properties as plastics. 'Plastic has climbed down, it is a household material. It is the first magical substance which consents to be prosaic' (Roland Barthes, *Mythologies*, Paladin, London 1973. First published in Paris, 1957). And where the craftsman working with traditional materials such as timber and metal has to do the best he can with what he has, the chemist, processor and designer can together fashion what they will from an infinite combination of chemicals. With god-like power they can create what was once a dream.

2
Transport

2.0 Introduction

In industrial societies, transport looks a better candidate than materials as the technological prime mover. It is surely not an outrageous oversimplification to ground much of American and European social and economic history in the second half of the nineteenth century in the diffusion of railway technology; nor then to see the automobile and its highways as the inheritor of the locomotive and its track as the multiplier of industrial growth, and the ringer of reverberating social changes, during the first half of the twentieth century.

The two readings in this chapter serve at least to qualify any attempt to derive modern life from its foremost means of locomotion. Villalon and Laux point out that the eventual prevalence of the petrol-engined car could scarcely have been foreseen in the 1890s, when it was outperformed by the light steam car. But even though the fortunes of the Locomobile company have much to do with the peculiarities of the American entrepreneurial culture, it does seem that in the long run, it was the technical superiority of the petrol car that was decisive. This is not in any case to speak of social effects. Joel Arthur Tarr reminds us that it was not the private automobile but rather 'public transit', above all the electric tram, which transformed the densely populated, nineteenth-century 'walking city' into the modern 'two-part' metropolis with business core and outwardly radiating suburbs. The simplicity of cause and effect here is alluring, but deceptive: the 'urban transportation crisis' was itself the result of arguably non-technological causes, such as population pressure and economic growth; and decisions to invest in public transport were taken within the framework of an American value system which combined a strong distaste for city life with a readiness to accept technical solutions to social problems. As Tarr implies, the appeal to a 'neutral' technology may well mask a desire for social control.

2.1 Joel A. Tarr, *From City to Suburb: the 'Moral' Influence of Transportation Technology*, 1973

> Humanity demands that men should have sunlight, fresh air, the sight of grass and trees. It demands these things for the man himself, and it demands them still more urgently for his wife and children. No child has a fair chance in the world who is condemned to grow up in the dirt and

---

Source: Alexander B. Callow, Jr, ed. *American Urban History: an Interpretive Reader with Commentaries*, New York: Oxford University Press, 1973, pp. 202–12.

confinement, the dreariness, ugliness and vice of the poorer quarters of a great city . . . There is, then, a permanent conflict between the needs of industry and the needs of humanity. Industry says men must aggregate. Humanity says they must not, or if they must, let it be only during working hours and let the necessity not extend to their wives and children. *It is the office of the city railways to reconcile these conflicting requirements.*
Charles Horton Cooley, 1891

The theme of the evil city and the virtuous countryside has persisted throughout American history. Warnings of the unhealthiness of the urban environment and the threat that the urban masses posed to American ideals emanated from public spokesmen and intellectuals throughout the nineteenth century. Articulators of this point of view, such as Thomas Jefferson, Ralph Waldo Emerson, and Josiah Strong, usually regarded rural life, in contrast to city life, as morally superior and generative of the virtues of health, strength of character, and individualism.[1]

Recently, however, Peter J. Schmitt and Scott Donaldson have expanded the classic urban-rural dichotomy.[2] As Schmitt observes, few of those in the late nineteenth and early twentieth centuries who attacked the city and praised their rural childhoods ever returned to farming. While they looked with nostalgia to the rural virtues, they were unwilling to sacrifice the opportunities for the acquisition of wealth found in the city. Rather than being imbued with the philosophy of 'agrarianism,' which demanded that man draw his livelihood from the soil, they settled for the 'spiritual values' of nature. And these values, which Schmitt calls 'arcadian,' thrived not only in the distant country, but also on the urban periphery in an area easily reached from the crowded city. In short, the suburb, *rus in urbe*, would enable Americans to pursue wealth and yet retain the amenities and values of rural life.[3]

The suburb, in view of those who praised it, had certain desirable characteristics that distinguished it from the city. While the city was crowded, dirty, and smoky, the suburb had an abundance of 'fresh air and clear sunlight, green foliage and God's blue sky.' In the suburbs, which were primarily residential, most people lived in single family dwellings or 'cottages' far from the smoke and noise of business and industry. Serenity and calm rather than hustle and bustle were the hallmarks of the suburbs. Natural surroundings, 'cottage' living, and peace and quiet provided ideal conditions for family life and for the raising of children, all within easy commuting distance of offices and factories in the core of the city.[4]

Without transportation technology, however, the suburb as the halfway house between city and country and as the embodiment of the best of these diverse worlds would have been impossible. A number of writers have commented at length on the significance of the concept of technology in American thought, but often they have viewed the machine as opposed to the values of the pastoral or rural ideal.[5] Many influential spokesmen on the urban scene in the late nineteenth and early twentieth centuries, however, viewed technology as putting arcadia within the reach of city dwellers who would otherwise have been denied its moral benefits. If suburbia was the garden that all urbanites should strive to reach, then it was the machine that made it possible to work in the city but live in that garden.

The 'machines' in this instance were various forms of urban transit ranging from the omnibus to the electric street railway, subway, and elevated railroad. All of these, theoretically, enabled the busy urbanite who labored in the city and resided in congested and unhealthy districts to continue to work in the city but to live in the suburbs in a superior environment. Statements about the role of transportation technology in permitting people to escape crowded and dirty cities occur throughout the literature on the city, but they appear with special frequency at the time of transit innovation. For, it was then that Americans, concerned with the social dangers posed by urban growth, reaffirmed their faith in technology and saw in transit innovation the means to escape successfully from urban problems.

Public transit developments had their largest impact on urban and suburban patterns roughly from 1840 to 1910. During these years urban population grew at a rapid rate, rising from 1,845,000 or 10.8 per cent of the total population to 44,639,989 or 45.7 per cent. European immigrants, mainly from the rural areas of Ireland, Germany, Austria-Hungary, Italy, Poland, and Russia, accounted for a large part of the new urbanites, a fact which increased the fears of nativists and their spokesmen that the cities threatened American values.[6] At this time, cities also greatly increased their size by annexing contiguous incorporated and unincorporated territory.[7] Within these burgeoning urban areas, public transit systems facilitated the dispersal of population, concentrated commercial activities, and reversed the spatial distribution of socio-economic classes as compared with the patterns of the pre-transit city, thus giving rise to the modern core-oriented metropolitan area.

This transformation of the city's ecological and demographic patterns had its beginnings in the 1840s and 1850s, when many large American cities faced what one historian calls an 'urban transportation crisis.'[8] Cities in the 1840s and 50s were still primarily walking or pedestrian cities characterized by a 'crowded compactness.' By 1850 New York had a population density of 135.6 persons per acre in its 'fully settled area,' Boston 82.7, Philadelphia, 80.0, and Pittsburgh, 68.4. Most urban residences were two- or three-story wooden or brick buildings. Those occupied by the working class and the poor were often packed with many families and their lodgers.[9] Land uses were not clearly specialized, and middle and upper class residences were interspersed with those of lower income groups and located comparatively close to manufacturing and commercial structures. In contrast to contemporary living patterns, the elite often lived close to the business and governmental center of the city, while the majority of workingmen distributed themselves in the outlying wards. Those workingmen who lived in the urban core packed into narrow alley dwellings, tenements, and cellars. As population pressure increased, already crowded lower class living areas disintegrated into slums.[10]

The growth of manufacturing and business activities accompanied the increased population congestion within cities. Most of the industrial and commercial development was clustered in the older sections of cities, especially near the waterfront seaports and river towns. These central locations offered savings in transportation costs as well as the benefits of agglomeration economies.[11] As the spatial needs of industries and businesses increased, residential population was pushed out of the central business areas. Many workers who were displaced moved into adjacent sections, creating increased problems and overcrowding and housing deterioration. These declining neighborhoods ringing the city center often attracted a large immigrant population who sought low-cost housing near their places of employment.[12]

The absence of a system of public transit exacerbated the problems of urban congestion. Without public transportation, persons whose workplace was separated from residence were forced to walk to work unless they could afford the expense of a private horse or carriage. The development of factories, banks, stock exchanges, and other such urban enterprises by the mid-nineteenth century greatly increased the number of persons confronted with a journey to work, often of some considerable distance. While the peripheral areas of cities as well as towns outside of cities grew rapidly in 1830s, 40s, and 50s, they were not necessarily bedroom communities. Towns close to the central city such as Lawrenceville and Birmingham near Pittsburgh and the Northern Liberties outside of Philadelphia had their own separate economic focus and did not serve as residential sections for a large number of persons employed in the central city.[13]

During the decades before the Civil War, however, several key transportation innovations occurred that eventually made possible a suburban life for many people employed in the central cities. The earliest of these innovations were the omnibus and the commuter railroad, both of which appeared in Boston, New York, and Philadelphia in the 1830s and in Baltimore and Pittsburgh by the early 1850s. The omnibus was usually drawn by two horses and carried about twelve to fifteen passengers over an established route of city streets for a fixed fare. A coachman, mounted on an elevated seat at the front of the vehicle, collected the fares and drove the horses, while the passengers, who entered through a door at the rear of the vehicle, sat on long seats along the side of the omnibus.[14] The steam railway, of course, was not originally intended for city-suburban travel, but was rapidly adapted to that use. Boston led the way in the development of commuter railroad traffic, followed by New York and Philadelphia. Regulations that prevented the running of steam locomotives on streets, however, restricted the use of commuter trains in the latter cities, as they did in Pittsburgh, for some years.[15]

The transit innovation that had thc most significant impact on urban patterns was the streetcar. First introduced in New York City in the early 1850s, it had spread to Boston, Philadelphia, Baltimore, Chicago, Cincinnati, and Pittsburgh by the end of the decade.[16] The running of cars on rails through city streets was a major technological breakthrough. Alexander Easton, the author of *A Practical Treatise on Street or Horse-Power Railways*, published in 1859, called streetcars the 'improvement of the age,' and in terms of the increased facility of the intra-city transportation of passengers, his enthusiasm was justified. Powered initially by horses and mules, then by cable, and ultimately by electricity, the streetcar dominated urban transit throughout the nation from the Civil War until the 1920s. Because of the much lower average fares of the streetcar as compared to the omnibus or commuter railroad, it had the greatest potential for enabling working people to move to residential areas with suburban characteristics.[17]

Although the electric streetcar, developed during the late 1880s, was often referred to as 'rapid transit,' this title properly belongs to the elevated train and to the subway, both of which operated on paths separate from street traffic and with trains of cars rather than single cars.[18] An elevated system was first constructed in New York in 1871, with the cars pulled by steam locomotives. Elevated systems with electricity as the motive power were developed in Chicago in 1892, Boston in 1894, and Philadelphia in 1905. Boston built the first subway in 1897, followed by New York in 1904, and Philadelphia in 1909. Elevated and subway trains traveled at a much faster rate of speed than did surface cars, but their high construction costs made them feasible only for larger cities with a high volume of traffic.[19] Public transit, therefore, for most cities, involved streetcar systems.

Without the implementation of these transportation innovations, American cities would have developed in a different spatial pattern. Public transit produced an urbanized area roughly characterized by a central business district (CBD) or downtown surrounded by concentric circles or zones with specialized residential and industrial functions. Traction lines radiated from the core into the residential areas. The CBD became, over time, a section devoted almost entirely to business and commercial uses with a concentration of office buildings, banks, specialized retail outlets, and department stores. The residential areas usually had sharply distinguished socio-economic patterns with the poorer sections near the core and the wealthier neighborhoods toward the fringe. Those areas with suburban characteristics (single-family detached homes and an absence of industry) developed both within and outside the city's boundaries. Many of the residents of those districts worked in the CBD and commuted by public transit; their daily journey-to-work significantly affected the tenor of life in the twentieth century city.[20]

The two-part city, divided between residential and commercial-industrial sections, developed over the last half of the nineteenth century and first decades of the twentieth century in response to public transit expansion. Heavily congested workingclass and immigrant living areas with problems of poor sanitation, high disease rates, and deteriorating housing, however, still persisted in large cities such as New York, Chicago, Boston, Philadelphia, and Pittsburgh. Manhattan's 10th ward, for instance, had a density of 523.6 persons per acre in 1890 and the 13th ward 428.8, with the population jammed into five and six story tenement houses. In Pittsburgh, in the same year, density in the wards surrounding the CBD had advanced to 121.9 persons per acre in a section with few dwellings over two or three stories.[21] Members of the middle and upper classes worried that these congested living conditions, especially among the poor and the alien, posed a danger to 'the moral integrity and the unity of the community.'[22]

Many commentators on the evils of urban life in the latter half of the nineteenth and the beginning of the twentieth century argued that the solution of the city's problems lay in the further extension of mass transit. They explained that through public transportation systems, men who labored in the city would be able to live and raise their families in the superior suburban environment. This theme was first articulated before the Civil War when omnibus and horsecar lines were begun and was heard with greater frequency toward the end of the nineteenth century as urban population growth and transportation development continued apace.[23] As more and more centrally employed middle class citizens moved to suburbs, urban spokesmen advocated the improvement and cheapening of mass transit to make possible a suburban life for the workingclass people remaining in the city.[24]

In the 1870s, for instance, Congregationalist minister Charles Loring Brace, in his book *The Dangerous Classes of New York*, warned of the deleterious effect of overcrowding on public morals and advocated the dispersal of population from city slums. Specifically Brace recommended the building of a subway or an elevated railway with cheap fares as the means to enable workers to settle in 'pleasant and healthy little suburban villages.' In the suburbs each family would have its 'own small house and garden,' while children would grow up 'under far better influences, moral and physical, than they could possibly enjoy in tenement-houses.'[25]

Other writers in the 1870s and 80s repeated Brace's arguments about the moral influence of improved transit but also went beyond the social context. William R. Martin in *The North American Review* and L. M. Haupt in *Proceedings of the Engineer's Club of Philadelphia* held that the building of rapid transit would have economic as well as social benefits. Rapid transit, they maintained, would encourage the development of manufacturing within the city, increase real estate values, and stimulate building in open areas. The solution of the city's congestion problems would therefore be accompanied by financial gain for the metropolis's businessmen and builders.[26]

Many cities, however, could not afford the expense of either an elevated rapid transit system or a subway, and their expansion appeared limited by the speed and capacities of the horsecar. The application of cable and electric power to street railways in the 1880s appeared to resolve this problem. During the ten years from 1880 to 1890, the length of street railway track in the United States jumped from 2,050 to 5,783 miles, an advance of 182 per cent at a time when urban population increased 56.7 per cent. Urbanites in 1890 averaged 111 rides per year, with totals over 270 rides per year per inhabitant in Kansas City, New York, and San Francisco.[27] In that year horses and mules still supplied the motive force on 71 per cent of the streetcar trackage, but this total plummeted during the decade. Most significant about the change in motive power was the increased speed of traction service. Streetcars now traveled at approximately 10 miles per hour, just about double the speed of the horsecar, greatly expanding the areas within commuting distance of the downtown core. From 1890 to 1902 track mileage increased from 5,783 to 22,577, almost all of it operated by electricity, and rides per urban inhabitant from 111 to 181. Five years later, in 1907, track mileage had jumped 53.5 per cent to 34,404 and rides per inhabitant to 250, as the use of public transit continued to far out-distance population increase.[28]

These transit developments gave further encouragement to those concerned with urban congestion that technology could solve the city's congestion problems. City boosters held this view as well as 'urbanologists.' In Pittsburgh, which began electrifying its streetcar system in the late 1880s, publications intended for visitors boasted of how the city's traction system permitted working people to live in 'cosy residences' in the suburbs away from the 'noise, smoke and dust of a great city.' Suburban life, in turn, said one Pittsburgh guide-book, prevented the 'breeding of vice and disease' and 'elevated in equal proportion the moral tone of the laboring classes.'[29]

The 1890 Census included, for the first time, a volume on transportation, with a special section on the 'Statistics of Street Railway Transportation' prepared by sociologist Charles H. Cooley.[30] A number of writers on urban trends in the 1890s used this material to demonstrate that the concentration of people in cities, with its deleterious effects, could be mitigated by improved urban transit. Carroll D. Wright, writing in the *Popular Science Monthly*, Thomas C. Clarke in *Scribner's Magazine*, Henry C. Fletcher in the *Forum*, and Cooley himself in his work on *The Theory of Transportation*, all agreed that while the conditions of modern industrial and commercial life necessitated concentration, 'humanity' required that men and their families live among 'sunlight, fresh air, grass and trees.'[31] In the words of United States Commissioner of Labor, Carroll D. Wright, adequate urban transportation was 'something more than a question of economics or of business convenience; it is a social and an ethical question as well.' For, concluded Wright, only suburbs could supply the 'sanitary localities, [the] moral and well-regulated communities where children can have all the advantages of church and school, of light and air . . .' so necessary to 'the improvement of the condition of the masses.'[32]

By the turn of the century, however, many commentators on urban congestion had come to question whether improved transit alone would make possible a suburban life for the families that filled the city's tenements. In his seminal work of 1899, *The Growth of Cities in the Nineteenth Century*, Adna Weber argued that the development of suburbs rather than other 'palliatives' such as model tenements, building laws, and housing inspection offered the best hope of escaping the evils of city life stemming from overcrowding. But while cheap and rapid transit was essential for suburban development, said Weber, it had to be accompanied by a shorter working day and inexpensive suburban homes if workingmen were to take advantage of an environment that combined 'the advantages of both city and country life.'[33]

Such sentiments were repeated by other urban reformers during the beginning of the twentieth century. Some, such as Benjamin C. Marsh, secretary of the Committee on Congestion of Population in New York, saw city planning, tax reform, and even municipal land ownership as indispensable accompaniments to improved transit if urban decentralization was to become a reality; others, such as Frederic C. Howe, believed that a single tax upon land values should accompany rapid transit development.[34] Many of the most heated local political battles of the Progressive Period were fought over the issue of municipal regulation or ownership of streetcar lines. Reform mayors such as Hazen Pingree of Detroit, Tom Johnson of Cleveland, and Samuel 'Golden Rule' Jones and Brand Whitlock of Toledo advocated better and cheaper traction service as a means to better the lot of urban workingmen; undoubtedly they believed that such improvements would permit residents of crowded inner city districts to move to neighborhoods where they could realize the suburban ideal.[35]

In its *Special Report on Street and Electric Railways* published in 1902, the Bureau of the Census of the Department of Commerce and Labor presented the most comprehensive statement yet on the street railway as a 'social factor.' The report observed that street railway development had come 'in response to an imperative social need,' and that urban transit had facilitated the dispersal of population from the city while encouraging the concentration of commercial and manufacturing establishments. Traction companies had also performed an important social service by transporting people from crowded cities to 'places of outdoor recreation.' But, added the report, because city transit service was often inadequate, it had actually hampered potential suburban growth. The report recommended increased speed, additional cars, and lower fares as one means to deal with transit deficiencies. Satisfactory suburban and interurban transit service in congested great cities, concluded the report, could only be supplied by elevated or subway systems in coordination with surface lines, but the expense of such systems perhaps precluded a 'wholly satisfactory solution of the problem of transportation in great and rapidly growing cities . . .'[36]

* * *

By the end of the first decade of the twentieth century, therefore, there existed a large body of literature which viewed urban transit as the means by which men could escape the evils of the crowded city and live in suburbs with the benefits of cottage living, clean air and sunshine, and close communication with nature. Some reformers, however, believed that changes such as tax reform, municipal ownership, and cheap housing had to accompany transit improvements if suburban living for the working-class was to become a reality. In reviewing this material, it is difficult not to conclude that many who advocated urban decentralization were as motivated by considerations of social control as by a desire to enable men to live more comfortable or healthy lives. City slums crowded with immigrants usually had high crime rates and poor sanitary conditions, and middle and upper class citizens worried about the threat of violence and disease posed by congested workingclass areas. Transit systems, by making it possible for working people to leave the unhealthy city for arcadian suburbs, seemingly offered a relatively inexpensive method of curtailing the threat of the slum.[37]

Were the advocates of a technological solution to the problems of the city stemming from congestion, whatever their motivation, misguided? As early as 1866, *The Nation* editorialized that the time and money required for commuting to suburbs eliminated it as an alternative for many members of the working-class, a conclusion also reached by the leading contemporary student of the beginnings of mass transit.[38] But what of later traction developments? Writing in 1890, journalist and social critic Jacob Riis pessimistically noted that rapid transit in New York had failed to resolve the problems of the tenement house slum. Technology had proved ineffective, he held, when faced by the 'system' resulting from a combination of 'public neglect and private greed.'[39] Even the 1902 Bureau of the Census traction report, which had praised the impact of the streetcar, noted that most suburbanites were well-to-do and that poverty and long work hours prevented many workers from utilizing public transit to escape the city.[40] And, in his study *Streetcar Suburbs*, historian Sam Bass Warner, Jr observes that urban transit provided a 'safe, sanitary environment' for only half of the Boston metropolitan population; the remainder were condemned to the crowded city. In addition, says Warner, the emphasis on a system of individualistic capitalism that promised a suburban life for those who could pay the price meant that the society neglected the immense social and housing needs of the workingclass and immigrant poor.[41]

Some of this criticism is justified. Many members of the working-class could afford neither the time nor the money required to commute to suburbs. Moreover, the stress on facilitating the movement of people from city to suburb via urban transit undoubtedly did divert attention from the housing and recreational needs of the poor who remained in the city. But this emphasis suited the American value system – a value system that combined a strong belief in the capacity of technology to solve social ills, a belief in the moral superiority of a suburban to a city existence, and a commitment to a system of private capitalism and decision-making which theoretically allowed each person to make his own choice of residence according to his income and preference.[42]

Given this set of values, the streetcar did not 'fail' in its promise. Urban transit systems did enable many citizens to leave the congested areas of central cities for living areas with more amenities. Housing vacated by these groups in turn provided more housing choice for those remaining.[43] That areas of some cities like New York and Philadelphia grew more rather than less congested during the streetcar era resulted as much from the constant influx of new urban residents as from deficiencies in the transit system. Given this rush into the cities, the ability of many of these newcomers to leave city congested areas as quickly as they did is perhaps more astonishing than the fact that congestion remained high in some wards.

Today the private automobile has replaced public transit as the chief means by which Americans commute between suburban residences and city jobs. Not surprisingly, many early auto boosters predicted that it would serve the same function that gave mass transit such an appeal generations earlier – the motor car would open a suburban existence for those who wanted to escape the crowded and unpleasant city.[44] The automobile has been largely successful in this role, and the flow of people leaving central cities for suburban amenities continues year after year. As the 1970 census revealed, more people live today in the suburban rings around central cities than in the cities themselves. The poor and the minorities, however, as during the streetcar era, still seem condemned to the central cities. The 'garden' continues to beckon, but obviously it will take more than transportation improvements alone to make possible a suburban life for all those who desire it.

Notes

1 For discussion of this literature, see David R. Weimer (ed.), *City and Country in America* (New York: Appleton-Century-Crofts, 1962), Morton and Lucia White, *The Intellectual versus the City* (Cambridge: Harvard University Press, 1962), and Anselm Strauss, *Images of the American City* (Glencoe: Free Press, 1961).

2 Scott Donaldson, *The Suburban Myth* (New York: Columbia University Press, 1969); Peter J. Schmitt, *Back to Nature: The Arcadian Myth in Urban America* (New York: Oxford University Press, 1969).

3 Donaldson, pp. 24–25; Schmitt, pp. xvii–xviii.
4 See, for example, Donaldson, pp. 24–27; Sam Bass Warner, Jr., *Streetcar Suburbs: The Process of Growth in Boston, 1870–1900* (Cambridge: Harvard University Press and the M.I.T. Press, 1962), pp. 11–14.
5 Leo Marx, *The Machine in the Garden* (New York: Oxford University Press, 1964); Marvin Fisher, *Workshops in the Wilderness* (New York: Oxford University Press, 1967); and Hugo A. Meier, 'American Technology and the Nineteenth-Century World,' *American Quarterly* (1958) X, pp. 116–130.
6 Charles N. Glaab and A. Theodore Brown, *A History of Urban America* (New York: Macmillan Co., 1967) pp. 25–27, 93–95, 107–111, 138–142.
7 Kenneth T. Jackson, 'Metropolitan Government Versus Suburban Autonomy: Politics on the Crabgrass Frontier,' in Kenneth T. Jackson and Stanley K. Schultz (eds.), *Cities in American History* (New York: Alfred A. Knopf, 1972), pp. 442–452.
8 George Rogers Taylor, 'The Beginnings of Mass Transportation in Urban America: Part I,' *The Smithsonian Journal of History* (Summer, 1966) I, p. 37.
9 *Ibid.*, pp. 37–38; Sam Bass Warner, Jr., *The Private City: Philadelphia in Three Periods of Its Growth* (Philadelphia: University of Pennsylvania Press, 1968), pp. 49–62; Warner, *Streetcar Suburbs*, pp. 15–21; Peter G. Goheen, *Victorian Toronto 1850–1900* (Chicago: University of Chicago Department of Geography Research Paper No. 127, 1970), pp. 8–9; Joel A. Tarr 'Transportation Innovation and Changing Spatial Patterns in Pittsburgh, 1850–1910,' (Pittsburgh: Carnegie-Mellon University Transportation Research Institute, 1971) pp. 2–5.
10 *Ibid.*, pp. 3–5; Goheen, pp. 8–9; Allan R. Pred, *The Spatial Dynamics of U.S. Urban-Industrial Growth, 1800–1914* (Cambridge: M.I.T. Press, 1966), pp. 196–197; David Ward, *Cities and Immigrants* (New York: Oxford University Press, 1971), pp. 105–109.
11 Taylor, p. 39; Ward, p. 86; Pred, pp. 148–152, 167–177.
12 Ward, pp. 85–109.
13 Taylor, p. 37.
14 Taylor, pp. 40–44.
15 Taylor, 'The Beginnings of Mass Transportation in Urban America, Part II,' *The Smithsonian Journal of History* (Autumn, 1966) I, pp. 31–38.
16 Taylor, 'Beginnings of Mass Transportation: Part II,' pp. 39–50.
17 Ward, pp. 125–143; George M. Smerk, 'The Streetcar: Shaper of American Cities,' *Traffic Quarterly* (Oct., 1967) XXI, pp. 569–584.
18 James Blaine Walker, *Fifty Years of Rapid Transit 1864–1917* (New York: Arno Press, 1970, reprint of 1918 edition), p. i.
19 Blake McKelvey, *The Urbanization of America 1860–1915* (New Brunswick: Rutgers University Press, 1963), pp. 76–85.
20 Smerk, pp. 569–584; Warner, *The Private City*, pp. 177–200; Warner, *Streetcar Suburbs, passim*; Ward, pp. 125–143. Cities obviously diverged in their patterns of development according to topographical factors and the individual locational decisions of businessmen and householders. I have not meant to endorse any particular theory of urban development but rather to point out that traction increased the tendency towards specialized residential and commercial districts. Some businesses and industries remained scattered throughout the city at the height of the streetcar period. See Raymond L. Fales and Leon N. Moses, 'Thünen, Weber and the Spatial Structure of the Nineteenth Century City,' to be published in Charles Leven (ed.), *Essays in Honor of Edgar M. Hoover.*
21 Roy Lubove, *The Progressives and the Slums* (Pittsburgh: University of Pittsburgh Press, 1962), p. 94; Adna F. Weber, *The Growth of Cities in the Nineteenth Century* (Ithaca: Cornell University Press, 1967, reprint of 1899 edition), pp. 460–464; Pittsburgh data from U.S. Department of Commerce and Labor, Bureau of the Census, *Vital and Social Statistics, Part II:*

Cities of 100,000 Inhabitants, Eleventh Census (Washington: Government Printing Office, 1895) XXI, pp. 293–311.

22 Lubove, pp. 10, 82.

23 For articulation of this theme in regard to the omnibus see Glen E. Holt, 'The Changing Perception of Urban Pathology: An Essay on the Development of Mass Transit in the United States,' in Jackson and Shultz, pp. 325–326.

24 Legislation providing for 'workingmen's' fares at a lower than usual cost was required for commuter railroads in Massachusetts in 1872 and for streetcars in Detroit in 1893. Parliament required commuter railroads in the London area to run 'workmen's trains' with low fares as early as 1860. For American developments see Charles J. Kennedy, 'Commuter Services in the Boston Area, 1835–1860,' *Business History Review* (Summer, 1962) XXXVI, pp. 169–170 and Melvin G. Holli, *Reform in Detroit: Hazen S. Pingree and Urban Politics* (New York: Oxford University Press, 1969), pp. 47–48; for London, T. C. Barker and Michael Robbins, *A History of London Transport* (London: George Allen & Unwin, 1963) I, pp. 173–174.

25 Charles Loring Brace, *The Dangerous Classes of New York* (New York: Wynkoop and Hallenbeck, 1880, 3rd ed.), pp. 57–60. For a study of Brace and his attitude toward the city see R. Richard Wohl, in 'The "Country Boy" Myth and Its Place in American Urban Culture: The Nineteenth-Century Contribution,' ed. by Moses Rischin, *Perspectives in American History* (Cambridge: Harvard University Press, 1969) III, pp. 107–121.

26 William R. Martin, 'The Financial Resources of New York,' *The North American Review* (Nov.–Dec., 1878) CXXVII, pp. 442–443; L.M. Haupt, 'Rapid Transit,' *Proceedings of the Engineer's Club of Philadelphia* (Aug., 1884) IV, pp. 135–148.

27 Charles H. Cooley, 'Statistics of Street Railway Transportation,' in Henry C. Adams (comp.), *Report on the Transportation Business in the United States, Part I, Transportation by Land, Eleventh Census* (Washington: Government Printing Office, 1895) XVIII, pp. 681–684.

28 U. S. Department of Commerce and Labor, Bureau of the Census, *Street and Electric Railways 1907* (Washington: Government Printing Office, 1910), p. 33.

29 *Illustrated Guide and Handbook of Pittsburgh and Allegheny* (Pittsburgh, 1887), p. 50; Consolidating Illustrating Company (comp.), *Pittsburgh of Today* (Pittsburgh: Consolidating Illustrating Co., 1896), p. 69; Knights Templar, *Official Souvenir 27th Triennial Conclave Knights Templar* (Pittsburgh, 1898), n.p.

30 Charles H. Cooley, 'Statistics of Street Railway Transportation,' in Henry C. Adams (comp.), *Report on the Transportation Business in the United States, Part I, Transportation by Land, Eleventh Census* (Washington: Government Printing Office, 1895) XVIII, pp. 681–791.

31 Carroll D. Wright, 'Rapid Transit. Lessons from the Census. VI,' *Popular Science Monthly* (Apr., 1892) XL, pp. 785–792; Thomas Curtis Clarke, 'Rapid Transit in Cities,' *Scribners Magazine* (May–June, 1892) XI, pp. 568–578, 743–758; Charles H. Cooley, 'The Social Significance of Street Railways,' *Publications of the American Economic Association* (1891) VI, pp. 71–73; Cooley, *The Theory of Transportation, Publications of the American Economic Association* (May, 1894) IX, no. 3; Henry J. Fletcher, 'The Drift of Population to Cities: Remedies,' *Forum* (Aug., 1895) XIX, pp. 737–745.

32 Wright, p. 790.

33 Weber, *The Growth of Cities*, pp. 467–475.

34 Frederic C. Howe, *The City: The Hope of Democracy* (Seattle: University of Washington Press, 1967, reprint of 1905 edition), pp. 202–206; Lubove, pp. 231–238. [. . .]

35 See, for example, Holli, pp. 33–55; Hoyt Landon Warner, *Progressivism in Ohio 1897–1917* (Columbus: Ohio State University Press for the Ohio Historical Society, 1964), *passim*; and Tom L. Johnson, *My Story*, ed. by

Elizabeth J. Hauser (Seattle: University of Washington Press, 1970), *passim*.

36 U. S. Department of Commerce and Labor, Bureau of the Census, *Street and Electric Railways 1902* (Washington: Government Printing Office, 1905), pp. 26–43.

37 On the theme of social control and urban decentralization see Lubove, pp. 131, 250–251.

38 'The Future of Great Cities,' *The Nation* (Feb. 22, 1866) II, p. 232; Taylor, 'The Beginnings of Mass Transportation: Part II,' pp. 51–52.

39 Jacob A. Riis, *How the Other Half Lives* (New York: Sagamore Press, Inc., 1957, reprint of 1890 edition), p. 2.

40 *Electric Railways 1902*, pp. 31–33.

41 Warner, pp. 160–161. Warner also feels that the suburbs failed to supply the reinforced community life that supposedly was a goal of those who left the city.

42 See *ibid.*, pp. 153–166.

43 Ward, pp. 120–121.

44 See, for example, Harlan Paul Douglass, *The Suburban Trend* (New York: Arno Press, 1970, reprint of 1925 edition); James J. Flink, *America Adopts the Automobile, 1895–1910* (Cambridge: M.I.T. Press, 1970), pp. 108–110; Blain A. Brownell, 'A Symbol of Modernity: Attitudes Toward the Automobile in Southern Cities in the 1920's,' *American Quarterly* XXIV (March, 1972), pp. 25–26.

2.2 L.J. Andrew Villalon and James M. Laux, *Steaming Through New England with Locomobile*, 1979

Comments about mechanized mass production as a unique American contribution to industrial development are as commonplace as they are misleading. Misleading because they ignore earlier European examples such as printing from movable type, shipbuilding in seventeenth-century Holland, the mechanization of textile manufacturing in Britain in the late eighteenth century, musket production in France about the same time, quantity output of pulley blocks for the Royal Navy during the Napoleonic Wars, and efforts in various nineteenth-century European cities at mechanized bread baking. The common feature in all these cases is that the manufacturer perceived a very large market for identical copies of the same product.

On the heels of these precedents, Americans seem to have done more than others with mechanized mass production in the nineteenth and twentieth centuries, especially with metal articles such as firearms, sewing machines, typewriters, and automobiles. To typify these achievements Henry Ford and the Model T, introduced in October 1908, are usually cited. One cannot deny the accomplishments of Ford, his associate James Couzens, and their engineers, but several other firms in the American automobile industry perceived a mass market and made thousands of cars for it before Ford did. The Buick 10, a small and cheap four-cylinder car offered first in 1907, found a big market before Ford's T. Olds from 1902 and Cadillac from 1903 produced thousands of cars annually while Ford was having problems starting his firm. Even earlier, the Electric Vehicle

Source: *Journal of Transport History*, vol. 5 (New Series), 1979, pp. 65–82.

Company had proposed to have 1,600 of its heavy and expensive electric cars built by the Pope bicycle and automobile factory in Hartford in 1899–1900. This scheme failed because the machines were faulty and the market wilted. But a demand for lightweight and inexpensive cars did exist. Neither Olds nor Buick nor Ford first demonstrated this in the United States; the Locomobile Company of America did, with a small steam car of which it sold some 5,200 from 1899 to 1903. By this performance Locomobile illuminated the possibilities and made it easier for the rest. Sprouted in the Boston region but rooted in Bridgeport, Locomobile also reminds us of the New England beginnings of the American automobile industry.

Early steamers in Boston

The Locomobile story begins in the late 1890s in the Boston region with three Yankee tinkerers: George E. Whitney and the twins F.O. and F.E. Stanley. George Eli Whitney was born about 1862, attended the Massachusetts Institute of Technology for a time, and then worked in machine shops and engaged in the manufacture of small marine engines in the 1880s and 1890s. He began building steam-powered carriages in East Boston in 1895–6.[1] His first weighed only about 700lb fully loaded and although later Whitney steamers were heavier this prototype set the standard for other New England steam cars – light in weight and simple in mechanism – in contrast to most European steamers which tended toward massiveness. Whitney made some half dozen more steamers in 1897–8 and organized the Whitney Motor Wagon Company in 1897 to manufacture them in quantity. This enterprise did not succeed. It appears that Whitney lacked both capital and production experience. A grand nephew of Eli Whitney of cotton gin fame, he cast himself in the role of inventor, not businessman. In 1899 he travelled to England and France with one of his cars, aiming to sell a licence to a manufacturer there. Many years later Whitney claimed he met an American car manufacturer in Paris and made a sale, but before pursuing this we must examine some other elements in the steam car story.

The Stanley brothers came from Maine where they were born in 1849. Perfectly fitting the stereotype of the eccentric Yankee inventor, F.E. Stanley in 1875 began to produce photographic dry plates in Lewiston. As this enterprise prospered F.O. Stanley joined the business and in 1888 they moved it closer to the market, locating it in the Boston suburb of Watertown, with a Newton post office address. (Many writers have mistakenly assumed that the factory itself was sited in Newton.) In 1896 F.E. Stanley saw an early automobile – it is not clear whether it was steam- or petrol-powered – and challenged himself to make a better one. He aimed at a weight of some 600lb so as to achieve rapid speeds and to use the flimsy bicycle-type pneumatic tyres of the day. But the Stanleys were not equipped to make many of the parts themselves and they had to buy most of them from specialist shops in the region. Their first steamer may have operated late in 1897. It raised little stir and the brothers went on to develop a new model. It used an engine manufactured by the Penny Machine Shop of Mechanic Falls, Maine, weighing just 32½lb. The steam boiler, frame, suspension, chain transmission, and general lines of the Stanley closely resembled the Whitney car but it weighed less, somewhat under 600lb loaded.[2]

When the Mechanics Fair in Boston opened in October 1898 it exhibited four automobiles. Two were electric, one offered by the Pope Company of Hartford, soon to become a major producer under the name of Electric Vehicle Company, and the other by Andrew L. Riker of New York City, ultimately a key figure in the present narrative. Another was a petrol-powered tricycle of which the French manufacturer, De Dion-Bouton, had sold thousands in Europe, and finally one operated by steam, but it was neither a Whitney nor a Stanley. Some days after the opening a Whitney carriage appeared. Then, for the parade on 9 November from the Mechanics Building to Charles River Park for races and other activities, F.E. Stanley showed up in his machine. He did not take part in the races but did demonstrate his car in an exhibition spin. His time for two miles, 5 minutes 19 seconds, was comparable to the 1,200lb Whitney and the French tricycle. The Stanley vehicle did win the hill climbing event, going up a 36½ per cent gradient without trouble.

The Stanley brothers decided to take advantage of public enthusiasm aroused by the exhibition and prepared to manufacture a batch of their steamers, perhaps 100 of them. The Mason Regulator Co. of Boston reported in December 1898 that it was making the engines for these cars and two months later the Stanleys said they had received orders for 85 cars. To undertake this new activity the Stanleys moved into a former bicycle factory adjacent to their photographic business in Watertown. The financial means and reputation as serious businessmen they had gained from the photographic enterprise appear as crucial elements in their success when contrasted with Whitney's failure to begin quantity production. They could order 100 sets of parts for a new and untried contraption and be taken seriously.

The Stanley steamer at this point employed a two-cylinder, 3½hp. engine with steam generated by a 14-inch boiler in which the heat rose through 298 small copper tubes. A petrol burner raised steam in this boiler. A delicate device, its fuel was kept under about 20lb pressure with a hand pump. The water tank held some 12–14 gallons, enough for a 15-mile trip. A chain transmitted power from the engine to a differential on the split rear axle. The frame was of bicycle tubing with a pair of elliptical springs in the rear and one transversely across the front. A tiller controlled the steering and the wire spoke bicycle-type wheels had two-inch tyres.

That the Stanleys late in 1898 undertook to manufacture up to 100 cars was almost unprecedented in the United States. There were only a few other widely scattered automobile makers at this point – in the mid-west Alexander Winton was the largest with an output of 22 in 1898 – but only the Pope Company made more than a few dozen (electric) cars in this year. Another basic decision of the Stanleys, to make small, lightweight, and inexpensive cars, also set an important pattern for the earliest period of American motor car manufacturing.

George Whitney also tried again to get into production, in 1899 licensing a firm located in Portland, Maine, and Lawrence, Mass., that made shoe manufacturing equipment to construct steamers on his designs. This move brought some confusion, however, because his licensee was the Stanley Manufacturing Company, operated by Frank F. Stanley. Although this concern did not make many cars and later they received the name McKay, two Stanley firms making similar steam cars at almost the same time in almost the same place do muddy the waters for the historian, not to mention contemporaries.

Walker and Barber

For a time, nevertheless, one of the Stanley companies left the automobile business and gave birth to two new firms, Locomobile and Mobile. This development of 1899 came largely through the efforts of two intriguing figures from America's great age of industrialization, the magazine magnate John Brisben Walker, publisher of the *Cosmopolitan*, and the asphalt paving king Amzi Lorenzo Barber. In F.O. Stanley's reminiscences he tells how the connection began one morning in February 1899 when he arrived at the steam car works in Watertown to find the New York magazine publisher waiting on the doorstep with an offer to buy into the company.[3]

John B. Walker, then 52 years old, had enjoyed a career fascinatingly varied and productive. Born near Pittsburgh in 1847, he attended a preparatory school in Washington, D.C., and when he was 16 at the height of the Civil War in 1863 he entered Georgetown University. Two years later, the war over, he went to West Point, a not unexpected move for a young man with military forebears on both sides of his family. In 1868 Walker withdrew from the Military Academy so as to accompany J. Ross Browne, the new U.S. Minister to China, where he remained for two years.

Returning to the United States in 1870, Walker entered iron manufacturing and other business ventures in West Virginia and quickly made the first of several fortunes, but the panic of 1873 wiped him out and ended this phase of Walker's business life. He then became a journalist, accepting an offer from the *Cincinnati Commercial* to write on economic issues. This led to a job as managing editor of the *Pittsburgh Telegraph*, and in a few weeks Walker went on to fill the same position with the *Washington Chronicle*. He remained with it for a few years until the paper folded in 1877.

Once again out of a job, John Brisben Walker took Horace Greeley's advice and headed west. He determined to apply irrigation to arid western lands. Eventually he bought up hundreds of acres near Denver and successfully raised alfalfa. More profitable were his speculations in Denver real estate and soon he amassed his second fortune. In 1889 Walker moved to New York and bought the ailing *Cosmopolitan Magazine*, a monthly begun three years earlier and owned for a short time by U.S. Grant, Jr. A 'great editor', Walker made the *Cosmopolitan* profitable and achieved a circulation of 300,000 by 1898. In 1895 he moved the entire publishing operation from downtown New York City to Irvington, a suburban community a few miles up the Hudson. The new *Cosmopolitan* buildings were designed by Stanford White, a socially prominent architect who later became the victim in the notorious Harry Thaw murder case. Walker loved to promote himself but he also engaged in a variety of public endeavours including a free correspondence school called the Cosmopolitan University.[4]

Alert to new opportunities, Walker took the motor car seriously from the beginning. *Cosmopolitan* publicized early automobiles and in 1896 he offered $3,000 to the car that would perform most creditably in a run of over 26 miles from New York's City Hall Park to Irvington. Although badly mismanaged, this affair, won by a Duryea petrol car, did generate the publicity that Walker wanted. Nearly three years later Walker decided to enter the automobile business himself. Apparently not interested in petrol power nor in the electric cars that did attract the interest and money of some New York capitalists, Walker opted for steam. In February 1899 he went up to the Boston area to see the Stanleys. When he offered to buy a half-interest in the fledgling enterprise they were astonished.

> Had he said he had come to buy a half-interest in our wives, I doubt if we would have been more surprised. We told him we didn't know as we had what might be called an automobile business, and we certainly did not want a partner; that we had difficulty enough in getting along with each other, and we did not want to increase our trouble by taking in a third party.[5]

Walker pressed his case, emphasizing his ability to publicize the steam car, but the twins could not be budged. He left, 'greatly disappointed'.

The more Walker thought about the idea the more he liked it and after two months he returned to Watertown, reappearing at the factory in mid-April with an offer to buy out the entire business. The Stanleys were not eager to sell so they set a figure they thought would send Walker away again, $250,000. 'Exactly the figure I had in mind, a quarter of a million', smoothly replied the publisher. The Stanleys signed, foreseeing an enormous profit, for they had invested less than $20,000 so far. Further negotiations provided that for $10,000 Walker received a ten-day option to buy the firm. F.O. Stanley quickly brought a steamer to Irvington, driving it from Watertown to Providence and taking ship from there to the metropolis, where he and Walker could demonstrate it in an effort to attract capitalists to supply the $250,000. Prospects included George and Edward Gould, sons of the famous Jay Gould, and William Rockefeller. But the financiers kept saying no. The option was about to expire when a man whom Walker had known in Washington over 20 years earlier and who now was a wealthy neighbour living just a mile down the river at Ardsley, Mr Amzi Lorenzo Barber, entered the picture.

A Vermonter, born in Saxton's River in 1843, Barber was the son of a forceful Congregational minister and spent most of his early years in Ohio. He entered Oberlin College at the age of 19 in 1862 and graduated five years later. After only a few months' post-graduate study at Oberlin's theological seminary, Barber encountered General O.O. Howard, head of the Freedmen's Bureau and currently developing the new university in Washington, D.C., named after himself and designed to provide higher educational opportunities for blacks. Barber joined Howard's new university and taught there for four years. His youthful idealism seems to have faded and shortly after his second marriage Barber left Howard in 1872 and entered the real estate business with some of his wife's relatives. Soon prosperous, in 1876 he witnessed an experiment that sharply changed his career. In Washington, the city authorities put down an asphalt pavement on Pennsylvania Avenue as a test venture. Grasping the possibilities of this innovation, Barber plunged into asphalt paving.

The Barber Asphalt Paving Company enjoyed gratifying success. By 1892 it had paved streets in some 30 cities and dominated the industry. A master stoke was Barber's conquest of the world's major source of natural asphalt, the Pitch Lake on the island of Trinidad. In 1888 he negotiated a lease with the British government to exploit the 114-acre lake for 42 years. In 1894 he tightened his grip on the supply side of the industry when he purchased the company controlling the Venezuelan supply of asphalt.

Now, in May 1899, Amzi Barber was Walker's last chance for financial support in his steam car venture. According to Stanley, who was driving, the asphalt king took his test ride and did not hesitate; he immediately decided to invest. He paid Walker $250,000 for a half-interest in the business, which made it possible for Walker to fulfil his commitment to the Stanley brothers. The new partners purchased from the brothers all steam car patents pending, the Watertown factory, and all automobiles in the course of manufacture. The Stanleys agreed to refrain from the independent production of steam cars before 1 May 1900, and furthermore signed on as general managers of the new company. Walker and Barber launched their firm on 17 June 1899, forming a West Virginia corporation named the Automobile Company of America, capitalized at $2½ million, with its office in New York City. Less than a month later the firm changed its name to Locomobile Company of America when it found that another concern had already pre-empted its original name. In October 1899 it raised its capital to $5 million. Public announcements indicated that Barber would serve as president, Walker as vice-president. In addition, a key figure entered the story, Samuel Todd Davis, Jr, Barber's son-in-law, who would function officially as treasurer.

Born in Washington, D.C., in February 1873, Davis attended the Columbian College Preparatory School there and in 1891 matriculated at Rensselaer Polytechnic Institute in Troy, New York, the oldest engineering school in the country after West Point. He graduated with a civil engineer's degree in 1895 and entered Barber's asphalt business. For a year Davis lived in Port-of-Spain as an executive of the Trinidad Asphalt Co. In 1897 he married Amzi Barber's daughter, Lorena. When Barber entered the infant automobile industry in 1899 his son-in-law came along enthusiastically.

Mobile

The first major steps the new automobile men took were away from each other. Barber and Walker split. Walker's plan had been to move the factory from Watertown to his own neighbourhood where he could watch affairs closely. In the spring of 1899 he negotiated to purchase a site at Kingsland Point in Tarrytown, a few miles north of Irvington. Walker, a self-made man who ran the *Cosmopolitan* as an absolute autocrat, had several sons whom he may have wanted to place somewhere in the management of the company. To Barber, another self-made man but less flamboyant, with a son and a son-in-law interested in the new venture, it may have looked like trouble to remain hitched to Walker and his large family with the factory in Walker's backyard. The issue came to a head shortly after 6 July 1899 when Walker closed the deal on the Kingsland Point property of 600 acres for $165,000. On 19 July the *Horseless Age* announced a Barber-Walker split with Barber keeping the Watertown factory and the Locomobile name while Walker retained the Kingsland Point estate with a new name, Mobile Company of America. Both companies had equal rights to use the Stanley patents and designs and the services of the Stanley brothers.[6]

Walker moved at breakneck speed to catch up with Locomobile, already in production at Watertown. On 14 July 1899 he broke ground for his new 'model' factory, designed by Stanford White to be 300ft by 50ft, and three stories high. Although the local builders Dinkel and Jewell rushed construction to finish in November, it took longer to find equipment, supplies, and to train labour. Only in March 1900 did Mobile turn out a completed automobile. Engines and wooden bodies had to be bought but most of the rest of the car was manufactured in this plant. Walker promoted his car vigorously and sold it at $650 through nearly 70 sales outlets.

Locomobile

Barber, like Walker, and several months ahead of him, moved quickly in 1899 to increase output of Locomobile's small steam cars, weighing only 400lb empty and priced at $600. As a first step he expanded operations in the Boston area. Manufacture of boilers and final assembly remained at the Watertown factory and Mason continued to supply the engines, but already in July he had obtained the Humber bicycle factory in Westborough (east of Worcester) to make the running gear. Then early in 1900 Locomobile leased the facilities of the Speirs Manufacturing Company in East Worcester and appointed its head, J.C. Speirs, superintendent of all Locomobile manufacturing operations. Speirs' company had once made bicycles. It now specialized in forgings and engines (replacing Mason) for Locomobile cars, although it is not clear if it had its own foundry.

Speirs took over his managerial post from the Stanley brothers, originally general managers but who faded quickly from the picture, for by September 1899 they had ended their connection with Locomobile. One can assume that they found it intolerable to work for other people. On specific issues they may have objected to the company's redesign of their car that was going to make it heavier, and they probably opposed Barber's fundamental decision to move the firm away from the Boston area.

Sometime in 1899 or 1900 George E. Whitney began to work for the Locomobile Company as a design engineer. Several decades later Whitney declared that he and Barber had met in Paris in 1898 and made some sort of financial arrangement. Whitney's date of 1898 could not have been correct but such a meeting may have occurred in August 1899 when both men were in Paris promoting their steam cars. Whitney did soon join the Locomobile staff although after the company left the Boston area he remained there until late in 1902. If Whitney did join Locomobile in the late summer of 1899 this appearance of a rival may have been another reason for the Stanleys to cut their ties with the firm.

Barber and his son-in-law Samuel Davis sought to concentrate production of Locomobiles in one place, and that place was not Boston but much closer to Barber's and the company's headquarters in New York City. They found the new location in Bridgeport, Connecticut, through the good offices of Dr I.D. Warner, of the Warner Brothers Corset

Co. of that city. They acquired 40 acres on Long Island Sound next to Seaside Park (most of whose land had come to the city from local hero P.T. Barnum), with railway and water transportation immediately available. Bridgeport, a city of 71,000 people, had several favourable features: metalworking and carriage-building shops to supply materials as well as skilled labour, and it was a convenient distance – 50 miles – from Manhattan by rail or water. The latter may have had a particular significance, for Barber was an ardent yachtsman. He could not wait for the new factory to be built before getting away from Boston so he found still another disused bicycle factory in Bridgeport and leased it as a temporary site for assembly of Locomobiles previously carried on at Watertown. Such assembly began in March 1900, after an estimated 600 of the machines had been built at the earlier location.

The new Bridgeport factory began to rise in November 1900, allowing some production activities to be transferred there from the Westborough and Worcester facilities in the following March. It was soon finished and Locomobile centralized all manufacturing in the new plant by June 1901, by which point nearly 3,000 of the cars had been sold. The company may have been turning out about 150 cars a month in its makeshift quarters. The main factory building, constructed of brick and steel, comprised two wings, 208 and 300ft long, making an L-shape, each wing 54ft wide.[7] It had four floors and a basement, supplying 127,250ft^2 of floor space. Smaller one-floor buildings housed a power plant, forge, testing shed and (water) pump house. Among early automobile factories the extent of Locomobile's floor space was second only to the De Dion-Bouton works in suburban Paris. Contemporaries considered the new factory a modern facility but we can see it as old-fashioned, reminiscent of New England textile mills with its long and narrow wings of four floors. Originally the steam engines in the power house drove machines in the other buildings by shafting but this outdated system was quickly abandoned in favour of generators to supply electric current for large motors to drive shafts and belts in each section of the factory. Presumably a separate motor for each machine was thought too large an investment. The company generated its own electric power because at this period the electric utility in Bridgeport supplied electricity to its customers for lighting only, and then only during the hours of darkness.

Descriptions of the factory reiterate that the company carried on most of the manufacturing processes in the plant. It did not have a foundry so it probably bought its engine castings from outside. Its wooden bodies and wire spoke wheels also came from outside suppliers. The tubular steel frames for Locomobiles were welded together in the forge shop and then whisked all the way to the fourth floor of the main building for painting, as were the wooden bodies. Boilers and burners were made on the first floor, water and fuel tanks on the third. Machine shops occupied parts of the first and second floors. Engine assembly and testing occurred on the third floor and final assembly on the second. This kind of factory arrangement was traditional, with different parts made in separate shops and moved from floor to floor on three lifts, ultimately to the final assembly operation where each car was put together by a group of men at a fixed station. One finds no attempt at continuous flow assembly, nor was this needed, it seems, for an output of 15 cars was considered a very good day's work and most days it fell far short of this figure.[8] All engines were thoroughly tested in the factory and after assembly the cars also received long test runs along the paths of Seaside Park. Such extensive testing suggests some difficulties in manufacturing and a lack of completely interchangeable parts.

Large-scale production

The Locomobile company from 1899 to 1901 was outstanding among all the other firms in the young industry for its large-scale production of low priced cars. Its production far outpaced the results of other domestic companies. The Electric Vehicle Company may have made as many as 1,500 of its Columbia electric cabs and passenger cars from 1899 to 1901 in Hartford but this effort with heavy and expensive machines had little success. This company's problems contrasted with the sales success of the light steam cars were instructive. Another object lesson, the small, French one-cylinder De Dion-Bouton petrol car introduced late in 1899, sold about 3,000 units up to 1902. Some of these (200?) were exported to the United States in these years, called Motorettes, and sold at $850 and up. The next large American producer, Ransom E. Olds, began making his successful Curved Dash one-cylinder petrol runabout in Detroit only in 1901, turning out some 425 in that year and some 2,500 in 1902. Its price began at $600, then rose to $650. In Kenosha Thomas B. Jeffery made 1,500 Rambler cars – also one-cylinder petrol runabouts – in 1902 at $750, and after these came Cadillac and Ford in Detroit in 1903 with production of 1,895 at $750–900 and up to 500 at $850, respectively.[9] These four large mid-western producers quickly followed the trail blazed primarily by Locomobile and Mobile which by 1901 had clearly demonstrated the reality of a large market for small, cheap cars.

Estimated Production of Locomobile Steam Cars

Year	Authors' estimates	Serial numbers
1899	600	1–1,200
1900	1,600	1,200–2,800
1901	1,400	2,800–4,200
1902	1,100	4,200–5,300
1903	500	5,300–5,800
Total	5,200	

Sources: Authors' figures are derived from the company's claims made in advertisements, copies of which are found in the Locomobile Collection, Bridgeport, and from *Cycle and Automobile Trade Journal* (August 1901), 14; *Horseless Age, x* (21 May 1902), 612; and *Automobile Topics* (12 July 1902), 354, (13 Sept. 1902), 1007. Serial numbers of the cars, compiled by Herb Ottaway and Hayden Taliaferro, are reported in *Horseless Carriage Gazette* (May–June 1977), 28. Company claims would be expected to magnify rather than minimize production; serial numbers assigned to a given year are not necessarily the same as actual production in that year.

Where did Locomobile find buyers for some 5,200 motor vehicles? The firm sought out clients with a vigorous marketing effort headed by J.A. Kingman, a college classmate of Samuel Davis, who emphasized the car's simplicity, reliability, and low price. Some buyers found that it was neither simple nor reliable, but with its low price it took advantage of the considerable publicity that had spread along with horseless vehicles since 1895. In comparison with petrol cars in 1900, when about 1,000 were manufactured in the United States, the Locomobile moved without apparent effort, with none of the vibration and little of the noise associated with that alternative. It had none of the problems of changing gears or worn out clutches. It could do wonders on short hills. Locomobile advertised heavily in a wide variety of publications, from *The Black Cat* and the programme of an international yacht race to the *North American Journal of Homeopathy*, the *Montreal Star*, *Savannah Morning News*, and *San Francisco Call*. It publicized its machines in auto shows, races, and stunts such as climbing Mount Washington in New Hampshire and Pike's Peak in Colorado. It established several branch sales agencies and recruited many dealers not just in its own region but from coast to coast. Late in 1899 it superseded its standard model no. 1 with an improved no. 2, which weighed 640lb empty and cost $750. This car replaced the lightweight Mason-built engine with a much sturdier one of its own (Whitney's?) design made in Speirs' factory in Worcester. Its water capacity was raised to 21 gallons, the steering tiller was moved to the side, and comfort and safety were increased by many other improvements.[10] Other models soon accompanied this new standard type for a total of six by March 1900, including a Locosurry weighing 800lb empty, seating four, and priced at $1,200, and a racing motorcycle with two saddles. Although the company continued to sell its cheapest car for $750 for as long as it made the steam type (except for a few months in 1901–2 when it sold at $850), most of its other models betrayed a tendency toward more substantial and higher priced cars.

Locomobile eagerly sought foreign markets. A few Canadians imported Locomobiles in 1899 and in October of that year some American bicycle interests formed the National Cycle and Automobile Company Ltd to make Locomobiles in Hamilton, Ontario. It began to assemble these cars and presumably made some of the components but in the summer of 1900 the Canada Cycle and Motor Company, owned by Canadians, bought out the National concern. It continued to make and sell Locomobiles until 1902.[11] Already in the summer of 1899 A.L. Barber and F.E. Stanley had visited Europe, arranging for Locomobile agents in London and Paris. While these steamers never caught on in France or the rest of the Continent, in Britain they found a receptive market. In the spring of 1900 one took part in the 1,000 Miles Trial, a reliability demonstration run by dozens of cars from London to Edinburgh and back. Several months later another Locomobile was driven from one end of Britain to the other, 981 miles, a journey that almost wore the machine out, and certainly tired the driver who had to pour a reported five tons of water into the water tank during the trip.[12] Nevertheless, a strong advertising campaign and the avid British demand for imported cars led to the sale of some 400 Locomobiles by 1903.[13]

Locomobile's sales of thousands of cars while other American firms had trouble making and selling dozens or hundreds actually disguised serious flaws and problems.[14] The 1899 models were flimsy, with frequent failures of bearings, sprockets and chains. Sturdier construction added weight which reduced the car's already low fuel economy. To heat its boiler it burned more petrol per mile than the typical petrol car of the period. To start the car when cold one had to warm up separately a petrol vaporizer or 'firing iron' to get the burner to operate, and then get up steam in the boiler. Ordinarily it took about half an hour to manage the entire process, and then there was always the chance that the wind might blow out the burner. While driving one had to watch the water gauge on the boiler and alter the rate of water feeding into the boiler when necessary. Miscalculations were common and burned out boilers occurred frequently. One also had to pump air into the petrol tank to keep its pressure up. Despite rumours, Locomobile boilers never blew up but some of the cars burned when the burners back-fired and set the car on fire. This particular feature made insurance companies reluctant to cover these cars. Boilers might fill up with scale and the cars could not be operated reliably in temperatures below freezing. One of the most glaring faults was high water consumption, about one gallon per mile. The water supply even for a 20-mile trip would weigh about 160lb. A possible solution was a closed system, sending the exhaust steam back to the water tank after running it through a condenser. The French steam cars made by Léon Serpollet and the American White steamers from 1902 employed condensers. But the engines used on the Stanley, Mobile, and Locomobile cars required copious lubrication and in a closed system this oil would foul the condenser and boiler. Locomobile did increase the size of its water tanks, a 1902 touring car carried 47 gallons, but the greater weight reduced the car's efficiency. Without redesigning the engine so as to adopt a condenser the only way to solve the water consumption problem would be to change the motive power itself, to shift to petrol engines.

New policies

Although we do not have complete sales figures for the Locomobile Company, these may have become somewhat discouraging late in 1901, after the opening of the new factory. Total production in 1901 probably failed to reach that of 1900, although the automobile industry nationally continued its rapid expansion. In this situation the company took three steps to protect itself. It prepared a new gamut of models for introduction in the spring of 1902, it filed suit against another steam car producer for patent infringement, and it decided to investigate the possibility of adding a petrol car to its line. These steps all accompanied the rise of Samuel Davis in the company, for he became president of Locomobile as 1902 opened, Amzi Barber becoming chairman of the board.

To investigate the petrol engine option Davis and Barber turned to a prominent electrical and automotive engineer, not a mid-westerner but a New Yorker, Andrew L. Riker. Now 33 years old, Riker was largely self-taught, although he had spent a year at Columbia College. In 1888 he had established the Riker Electric Motor Company and in the mid-1890s began making a few electric cars. In June 1899 he formed the Riker Electric Vehicle Company operating from Elizabethport, N.J. Riker sold this concern to the Electric Vehicle Company of Hartford in December 1900, becoming vice-president and superintendent of the larger firm. Already the Hartford company was in serious technical and financial trouble and Riker, apparently persuaded by his own long experience that electrical power was not the answer, began experimenting with petrol cars. In August 1901 Electric Vehicle alerted its agents that it had in preparation some two- and four-cylinder petrol machines for delivery in October. These announcements were premature and may have been issued by Riker without the assent of President George Day, ill and away from his office during much of 1901. In any event Riker did drive a new two-cylinder petrol model in the September 1901 New York to Buffalo Endurance Run. Designed not as a horseless buggy but on the French style developed by Emile Levassor, it had its engine with two vertical cylinders in front. This car had difficulties in the Run and Riker withdrew from the competition north of Albany. In December 1901 he also withdrew from the Electric Vehicle Co. which did not pursue development of his petrol machines.

No matter, Riker had known Amzi Barber and Samuel Davis for several years through participation in races and on automobile committee work. When he was 27 years old Davis had been elected the first president of the National Association of Automobile Manufacturers in December 1900 and Riker sat on its board of directors. They got together quickly and Andrew Riker joined Locomobile in January 1902 as the company agreed to finance his further development of a two-cylinder petrol car; a decision to manufacture such a machine might come later. To protect current sales of its steam cars Locomobile tried to keep Riker's activities quiet and he carried them on at the Overman Automobile Company in Chicopee, Mass. This firm had earlier made some experimental petrol cars and were currently producing the Victor steam car in small numbers. Locomobile made an unpublicized alliance with Overman in January 1902.

The interest in petrol cars and in Andrew Riker did not at all mean that Locomobile had chosen to abandon steam. At the very time that Riker was beginning his duties with Locomobile the firm acted vigorously to protect its steam car business by bringing suit against the Stanley brothers for patent infringement. The Stanleys had left Locomobile in September 1899 but they continued their experiments with steam cars. By early 1901 they bought back their property and equipment in Watertown from Locomobile for just $20,000 and they offered a new and improved steam car for sale in the winter of 1901–2. They hit a snag, however, for Locomobile now controlled George Whitney's patents and it claimed that the new Stanley infringed on one of them, granted to Whitney on 3 July 1900, for which he had made application as early as 30 April 1897. When an amicable agreement between Locomobile and the Stanleys could not be reached, Locomobile filed suit in January 1902. Later in the year it brought similar suits against the White Company of Cleveland and other steam car manufacturers. As it turned out, this affair did not grow into another Selden case which had so exercised the petrol car makers. The Stanleys solved their problem by thoroughly redesigning their car – most notably they adopted transmission by gears rather than chain – and eventually they bought all the steam car patents owned by Locomobile. The Stanleys made steam cars for many years thereafter, usually about 500 or 600 per year, gradually improving the design and adding weight, finally adopting a condenser in 1915. The company went out of business in 1924 after having produced some 14,000 steam cars since 1901.[15]

To appeal to the market Locomobile had to do more than enforce its patent rights. In the early spring of 1902 it began introducing new model steam cars to replace the types it had manufactured for about two years. In its most striking innovation it almost completely abandoned the earlier buggy design and on all but one of the seven 1902 models placed some of the mechanism under a box on the front end of the car, in this way imitating the best European petrol car designs. Artillery wheels with 3½-inch tyres were mounted on most of the new models and on some of them buyers could specify a steering wheel rather than a tiller. Water tanks held from 30 to 47 gallons and prices ranged from $750 to $2,400. Locomobile offered 'a car for every pocketbook'. It also offered steam car components, including engines, to other manufacturers.

By August 1902 the company also found Riker's petrol car worthy of adoption, decided to manufacture it, and made Riker a vice-president with 1,000 shares of Locomobile stock. Within weeks the public learned the news. *Automobile Topics* reported in late September that Locomobile soon would offer two- and four-cylinder petrol cars designed especially for touring (where the steamers' thirst for water often proved an embarrassment). Riker had designed heavy and expensive cars, about 2,000lb in weight with a $5,000 price tag on the four-cylinder model. Deliveries began in November.

Steam in trouble

By the end of the summer of 1902 the petrol car may have become not only another product for Locomobile but a possible way out of financial difficulties. The market for steam cars, at least those made by Mobile and Locomobile, was evaporating. Looking back some 13 years later, J.B. Walker of Mobile summed up the two major problems of this type of car as the 'necessity of filling its water tanks every so many miles – fifteen or twenty – with the regularity that one must water a horse. The other was the constant danger to the careless driver of having his boiler burnt out, through neglect in turning on his pumps at the proper time.'[16] As these negative features became known to prospective buyers many turned to other makes. Those still interested in steam cars could buy a better one in the Cleveland-built White (introduced in 1901) with a semi-flash boiler and condenser, or one of the new cars made by the Stanley brothers. Those considering petrol could now choose from the low-priced models offered by Olds (from late 1901), Rambler (1902) and soon a variety of others. One of Olds' first advertising slogans alluded to the complexity of the steamers: 'Nothing to watch but the road.' A correspondent to the *Horseless Age* revealed the disaffection from Locomobile and Mobile cars by 1903 when he remarked that 1900 or 1901 models of these types could be acquired for $100 or $200.[17]

After some consideration of a possible shift to petrol cars the Mobile Company gave up, closing its doors early in 1903 after selling up to 600 steam cars altogether from 1900. Within a year Maxwell-Briscoe bought the Tarrytown plant to make petrol cars, in turn selling it to Chevrolet in 1915. John B. Walker also sold the *Cosmopolitan*, to Hearst in 1905, and a few years later retired to alfalfa farming in Colorado where he lived on until 1931. Walker lost a sizeable sum of money with his gamble on steam, but Barber apparently sustained an even greater loss.

The Locomobile Company operated in the red for the year ending 31 August 1902 and the situation grew worse. In March 1903 Amzi Barber's money and credit ran out. Locomobile earnings did not appear to have offset his large outlays for designs and patents to the Stanleys and to Whitney and for the new Bridgeport factory. Other personal expenditures may have aggravated his position. In sum, he needed financial relief. He appealed to his brother-in-law J.J. Albright of Buffalo, with whom he had once been associated in Washington, D.C., real estate development. In March 1903 Albright agreed to endorse some Locomobile commercial loans in return for collateral comprising various of Barber's securities and properties. From this point Barber seems to have dropped out of active Locomobile management and in 1904 he returned to the asphalt business until his death in 1909. As Barber gave up the moneyman's role Albright took it on and became the leading stockholder in 1910. Actual management of Locomobile devolved upon S.T. Davis, Jr and Andrew Riker, with the latter in charge of engineering and production. In 1903 the company disposed of an estimated 500 steam cars and just 77 of the new petrol type. It lost money again, but in 1904 production of petrol models rose to 199 with steamers down to a handful; the company declared a profit.[18] The steam era had ended for Locomobile.

Locomobile's later history

To gain publicity and sales for its petrol cars Locomobile raced them, following the line of many European makers of powerful and expensive machines. In 1908 it gained a lustrous triumph when one of its racers won the Vanderbilt Cup Race on Long Island, the first domestic winner of the premier American auto race. This victory and healthy sales in these years repaired the company's finances. In 1912 it entered the truck market with a large 5-ton model. This branch of production grew to major proportions once the European war broke out. Britain, France, and Russia bought Locomobile trucks and in 1916 the United States Army began using them on the Mexican expedition. In this year the company gave the name Riker to its trucks (Davis had died suddenly in 1915, a serious blow) so as to preserve the Locomobile name for its fashionable and non-commercial passenger cars. Output of cars had to be cut in 1917 and 1918 to concentrate on trucks, as well as to prepare for manufacture of Liberty aircraft engines and to work on a heavy tank.

The success and profits Locomobile enjoyed during the war did not continue long into the post-war period. In 1920 the company joined the Mercer and Simplex automobile companies in a new group organized by Emlen S. Hare, a former Packard executive, and called Hare's Motors. Hare's vision was to manufacture quality cars in quantity and sell them at substantially lower prices than hitherto. Whatever the merits of this approach, it had little chance of success, for the post-war recession of 1920–1 set in and the market for Hare's Locomobiles and Mercers faded away. The industrial group collapsed and a grievously weakened Locomobile regained its independence in August 1921. As its business did not improve

Locomobile went into bankruptcy in 1922. William Crapo Durant, now building a new automobile empire after being ousted from General Motors for the second time, bought Locomobile for only $1,750,000. This sum allowed about 35 cents on the dollar for the company's creditors.

The Bridgeport factory resumed production but never did well. Durant directed it to make some medium-priced cars as well as expensive models but used the Locomobile name on both. The factory also turned out parts for other Durant cars. Death throes began early in 1929. By march Locomobile production had stopped. The management struggled fitfully, engaging to overhaul Paramount taxicabs during the summer of 1929 while preparing a new Locomobile model for the fall, but the Crash then erased this dream and put a final end to Locomobile.

Conclusion

However impressive Locomobile's elegant petrol cars may have been, in the sweep of automobile history in the United States its little steam cars of 1899–1903 had a much greater impact. The thousands of these models that buyers snapped up showed the reality of a large market for cars despite the wretched roads that inhibited inter-city driving. Their low price, in the $600–$850 range, put them within the means of ordinary professional men and the middle class generally. Other manufacturers soon moved into this market on the heels of Locomobile with thousands of small petrol cars and the American automobile boom was under way. We have found no evidence that the Locomobile added anything significant to production methods despite its relatively high output. Its production was not large enough to require innovations in methods so as to avoid heavy investment in new plant or to overcome shortages of skilled labour.

Walker and Barber chose a steam-driven car when they entered the business. This seemed wise enough in 1899, when American petrol cars appeared crude and unreliable. In France, the leading area of automobile development at this time, steam already had lost out to internal combustion but most Americans did not know this and those who did may have considered the French precedent irrelevant. In a few years Americans did find that steam propulsion, given the current state of technology, could not do as much and caused more difficulties than petrol engines. A more careful design of the early Locomobile to make it sturdier, simpler, and safer to operate might have enabled it to survive longer but Walker and Barber at first kept the original Stanley design on the market. The new Locomobile models that began to appear at the end of 1899 corrected or mitigated some of the earlier faults but improvement seemed to come faster in the competing petrol cars. By 1902–3 this type's advantages even converted Locomobile's management.

When Locomobile turned on to the same road as its mid-western rivals and switched to petrol it completely abandoned the low-priced field where it had disposed of so many cars through an extensive dealer network. It chose to appeal to the wealthy with elegance, quality, and high price. It seems as if Davis and Riker, young engineers, preferred a strategy emphasizing technical excellence as against the policy of Walker and Barber, older and experienced businessmen, who had tried to appeal to a mass market. It is possible that if Locomobile had continued in the low-priced market with its petrol cars it might have competed successfully with the mid-western companies and kept New England as a centre of this industry. However, its decisions and the earlier one of the Pope Company of Hartford to go to electric cars, a dead end at this time, meant that the industry would shift westward.

Notes

1 A recent survey of steam car history can be found in G. Levine, *The Car Solution: The Steam Engine Comes of Age* (New York, 1974).
2 *Horseless Age*, III (Oct. 1898), 44; *American Machinist*, XXI (1 Dec, 1898), 897–8; F. O. Stanley quoted in T. Derr, *The Modern Steam Car* (Newton, Mass., 1932), 45–6.
3 Derr, *op. cit.*, 48.
4 F.L. Mott, *History of American Magazines*, 4 vols. (Cambridge, Mass., 1930–57), IV: 35, 480–91.
5 F.O. Stanley quoted in Derr, *op. cit.*, 48.
6 *Horseless Age,* IV (19 July 1899), 12.
7 This description is drawn from City of Bridgeport, Building Permits 6094, 6161, 6162, 6163, 6234; *Automotor Journal* (Oct. 1901), 21; *Cycle and Automobile Trade Journal* (August 1901), 14–16; and *Automobile Topics* (August 1901), 563–70.
8 Annual production in 1901 is estimated at 1,400, or an average of under five per day.
9 The figures are stated or suggested by W. Greenleaf, *Monopoly on Wheels* (Detroit, 1961), 65–74; J. B. Rae, *American Automobile Manufacturers* (Philadelphia, 1959), 68–70; J. M. Laux, *In First Gear* (Montreal, 1976), 30, 212; G. S. May, *R. E. Olds* (Grand Rapids, 1977), 136, 187; T. Mahoney, *The Story of George Romney* (New York, 1960), 255; G. Brigham, *Serial Numbers of the First Fifty Years*, 5th ed (Marietta, Ga, 1974), 93; W. C. Leland and M. D. Millbrook, *Master of Precision* (Detroit, 1966), 69, 72; A. Nevins, *Ford*, 3 vols (New York, 1954–63), I, 231, 240–1, 246, 288.
10 C. Figge, in *Horseless Carriage Gazette* (May–June 1977), 29–31.
11 H. Durnford and G. Baechler, *Cars of Canada* (Toronto, 1973), 52, 84, 230–1.
12 Lord Montagu of Beaulieu and A. Bird, *Steam Cars: 1770–1970* (New York, 1971), 95.
13 M. Sedgwick in G. N. Georgano (ed.), *Complete Encyclopedia of Motor Cars* (1968), 349.
14 An excellent account of the technical aspects of the Locomobile is in Montagu and Bird, *op. cit.*, 96–107; issues of the *Horseless Age* in 1901 and 1902 frequently carry letters and articles on steam car problems and advantages.
15 See C.C. McLaughlin, 'The Stanley steamer: a study in unsuccessful innovation', *Explorations in Entrepreneurial History,* VII (1954), 37–47; and R.C. Sprague, Jr, 'The Stanley steam cars', *Antique Automobile* (Jan.–Feb. 1977), 30–41.
16 *Motor* (Jan. 1916), 73.
17 *Horseless Age,* XII (8 July 1903), 46.
18 Brigham, *op. cit.*, 59; Locomobile Collection, Bridgeport Public Library.

3
Communications

3.0 Introduction

One of the most voguish forms of technological determinism in the 1960s was the thesis underlying Marshall McLuhan's images of an 'electric age' and a 'global village': that society is shaped more by the technology of its communications media than by the information it carries. Less audacious communications theorists have traced to the original electrical medium, the telegraph, many constitutive features of modern society: the standardization of time, the language of journalism, the internationalization of commodity markets, and ultimately, the whole of modern consumer culture.

The successive effects in everyday life of the telephone, the gramophone, cinema and radio could doubtless be amplified in similar ways, but we need reminding that each of these innovations in communication has complex social origins. The two readings in this chapter serve as that reminder, in their contrasting styles. Michael Chanan, a radical filmmaker and historian of the film industry, surveys the range of nineteenth-century communications from a politically committed standpoint on the relations of technology and society. Despite its comparative tailpiece, Claude S. Fischer's main focus is considerably finer, though he drills correspondingly deeper into the primary source bedrock of the social history of technology, and still manages to offer generalizations on the causal relations of technological innovation and social change in some ways as fundamental as those of Chanan.

Chanan holds that the causal relations of communications innovations and socioeconomic change are mutually reinforcing. There is a corresponding ambivalence in the social significance of the different media he discusses: they are both servants of the capitalist and professional classes, and their instruments of domination; both means of consumption and means of production. They are the condition of production of the individual entrepreneur, as well as the product of 'the whole ensemble of individual enterprises.' Although in Chanan's scenario the dominant protagonists are the capitalist enterprise and a supportive state, these are often seen to be slower on the uptake than amateur and 'alternative' users. Fischer's article provides a striking example. The fact that in 1920 a greater proportion of American farms had telephones than did households in general, turns out to be completely contrary to the corporate strategy of the dominant Bell Company and its subsidiaries, Western Electric and AT & T.

Chanan's and Fischer's contrasting studies make the same point: that human agency, whether of dominant social classes or of ordinary consumers, is decisive in determining the effects of modern science-based technologies in everyday life. So far is any simplistic notion of 'autonomous technology' undermined.

3.1 Michael Chanan, *The Reuters Factor*, 1985

The year is 1848. A young German Jewish intellectual involved in radical journalism is obliged to leave his homeland, and heads for Paris. He is the third son of the Provisional Rabbi of Cassel. At the beginning of the decade he had settled in Berlin, where he was baptised and married the daughter of a banker. Assisted by his father-in-law's capital, he bought a share in an established bookshop and publishing business which then, under his guidance, brought out a number of 'democratic' pamphlets in the year of the Revolution.

In Paris, he joins the staff of a news agency owned by another Jew of letters, Charles Havas; he works as a translator. A year later he leaves his employer and sets up his own rival news-sheet. It fails, and he moves to Aachen, where, on 1st October 1849, Europe's first commercial telegraph line opens: the Prussian State Telegraph line from Berlin. He sets up in business supplying local clients with the news from the Prussian capital, soon expanding to supply clients in Antwerp and Brussels. When the French open a line the following spring from Brussels to Paris, he bridges the gap – first with carrier pigeons, and then with horses. But competition is fierce. In Berlin, another Jewish ex-Havas employee, Bernhard Wolff by name, has set up an agency with the backing of the electrical entrepreneur Werner Siemens. One day Siemens meets our hero, and advises him to go and start a cable agency in London. Born Samuel Levi Josaphat, he is known to history by his baptised name, Paul Julius Reuter.

I

One of the myths that has built up around high technology is the vision of a completely wired-up society. Everyone's home is to be wired in through their television sets and computers to other computers, which are wired in, in turn, to still more computers, and everyone and everything is fully programmed. You've not just got all the home entertainment you could want, you're also in instant two-way communication with the whole world. You've no need to go out shopping because you can order things and even pay for them through your TV set, and then consult your bank manager at the touch of a button. You'll even – they tell you – be able to work from home. You'll never need to step outside your front door, but if you want to, you'll have portable extensions to all your devices, so you need never be without their wonderful convenience.

This is the vision which is promoted by the government of Mrs Thatcher. These people see everything linked in together in a comprehensive system, designed to distribute pre-packaged dollops of entertainment and 'information' on a strictly commercial commodity basis: a 'free market' which they know perfectly well is loaded in favour of authority. [. . .]

Source: Michael Chanan, 'The Reuters Factor. Myths and realities of communicology: a scenario' in *Radical Science*[16], *Making Waves: The Politics of Communication*, London: Free Association Books, 1985, pp. 109–134.

They must know (but never let on if they do) that there will be masses of people excluded from participation in this heaven on earth. They are therefore equally set on devising ways of keeping these masses under control, and of nipping rebellion in the bud. Those who warn about the technology of social control which is now being introduced into the police force are right to do so. The police state and the information society are constructed on the same foundations and by the same means. This much is probably already apparent to anyone who reads the newspapers and watches television, for all their disinformation. But the issues this process raises are a minefield, and many disturbing notions seep through the public discourse. One paradox is that a welter of articles and programmes proclaim the new technology as a means of what the hacks have the nerve to call 'liberation' and 'democratization' – both an end to the drudgery of work and a new promise of instant participation; and yet in Britain in 1984, the real accompaniment to all the talk of an astonishing future is the famous return to Victorian morality of the Thatcherites. Very well, then, let's go back to the nineteenth century, and look at where the new technology started.

II

The apologists are right about one thing. The age of entrepreneurial capitalism didn't just consist of fortunes made from new industrial processes; it also saw the birth of modern communications. Communications are an integral part of the capitalist mode of production. First there's the development of the physical conditions: as goods came to be distributed in distant markets and foreign raw materials were increasingly employed in production, especially those that could be cheaply extracted from the expanding colonies, the improvement of terrestrial transport, which in feudal society was unorganized, became imperative. In due course, 'the feverish haste of production, its enormous extent, its constant flinging of capital and labour from one sphere of production into another, and its newly-created connections with the markets of the whole world' (Marx, p. 384), also make improved communications imperative for another reason: it creates a generalized need for the dissemination of a new type of information, consisting in stock market quotations, raw materials prices, credit rates, statistics and news.

In short, by the early nineteenth century the growth of capitalism had created pressing needs for improved commercial intelligence. The two went hand in hand. Between 1800 and 1913, as modern communications were developing, the value of world trade expanding (according to one estimate) more than twenty-five fold, from £320m to £8360m. The relationship between the two is an aspect of what we can call the Reuters Factor, which functions like a multiplier that turns an increase in the supply of information into an increase in business.

Big banking houses like Rothschild's started off with their own communications systems, using couriers and carrier pigeons. Newspapers like *The Times* set up networks to provide regular information on market prices in different financial centres, and Havas in France made the reputation upon which he built his news agency by supplying the Bourse with the European exchange rates. On the day that the Prussians opened a state telegraph line between Berlin and Aachen, 1st October 1849, two new contenders, both of whom started with Havas, set up in business: Bernhard Wolff in Berlin and, at the other end, with an overland link to Paris, Paul Julius Reuter. As the telegraph network was extended, Reuter preceded it, and in 1851, he established himself in London in order to exploit the submarine link between Calais and Dover.

The process which is described in the present essay is coextensive with the trajectory which Reuters has followed since, from its inception to the present day, when it's become a success story in the application of the latest communications technologies. Although, as we saw, it begins as a conventional business undertaking, Reuters' trajectory has been far from conventionally typical of capitalist growth (what is, when you come down to it?). But its very singularities make it typical in a kind of unconventional way: its discovery that news is a very peculiar kind of commodity, the question of its relationship to the authority of the State, its changing structure of ownership, from entrepreneurial to corporate. All these things, upon examination, draw our attention to internal features of capitalism which are frequently invisible. We find we're dealing with the bloodstream of the system: the flow of information upon which the health of the body depends. News is only part of this.

Reuter realized very early that the supply of news to newspapers by itself could not succeed in generating a profit. For one thing, the newspapers were jealous of their own prerogatives, especially, in London, *The Times*. But the Stock Exchange was a different matter, and Reuters' first English clients were private commercial subscribers. From biographical information (Storey), it is clear that Reuter understood something of the ideological relation between news and commercial information; he knew, for example, that he couldn't compete with the newspaper correspondents' form of comment, interpretation and graphic description. Telegraphy was too expensive. So he emphasised the telegraph's advantages of speed and conciseness, and constructed a model of reporting 'facts' which came from the criteria of commercial usage, where it was pretty unambiguous – the facts had numbers attached to them. He also talked about objectivity – which was pure expediency. He was a foreigner reporting imperialist wars and diplomacy, so he couldn't afford to incur the displeasure of Her Majesty's Government. He was also trying to satisfy the varying political and ideological inclinations of different newspapers.

To serve the interests of Empire and of them all, Reuter devised a code of practices for his establishment which succeeded in meeting all prejudices. It has essentially changed very little since it emerged in the 1860s and 70s, and lies at the basis of modern bourgeois ideologies of journalism. This is hardly accidental. Reuters made itself essential to the growing press, especially the growing number of provincial newspapers, unable to afford their own foreign news reporting services, who came together in the Press Association in 1865. The relationship was sealed in an agreement by which the Press Association provided the information for Reuters' cables to overseas clients. (The partnership strengthened Reuters internationally too, and this was the period when the big international agencies were carving up the world among them.) Subsequently, the Press Association were to join the Fleet Street newspaper proprietors in corporate ownership of Reuters, to preserve its independence.

For all this, news for the press alone was no way to make the venture a success. Periodically, throughout its history, Reuters has made investments in the improvement of its basic activities in commercial intelligence. They include especially the acquisition in 1943 of Comtelburo, a commercial and financial service designed to meet the needs mainly of banks and commercial institutions. [. . .]·

III

The new telegraphic news of the nineteenth century played an enormous role in shaping the sensibilities of the new and evolving industrial press. The total circulation of daily papers in the UK in 1854 was less than 100,000, of which *The Times* claimed 51,000. Sixteen years later, during the Franco-Prussian War, The *Daily News* alone had 150,000, and the editor of *The Times*, Mowbray Morris, instructed his correspondents abroad that the telegraph had superseded the newsletter and had rendered necessary a different style and treatment of public subjects. These sixteen years cover a tale of resistance by *The Times* to Reuters' growing monopoly in the provision of foreign news to the country's press. But not only was Reuters zealous about its special reputation for accuracy and trustworthiness, *The Times* was finally forced to give in when Reuters also proved more efficient and resourceful in the process of news gathering itself.

One effect of telegraphic reporting, due to the rapidity of transmission, was to establish a daily 24-hour cycle in the production of the news. Indeed, the new simultaneity of events threatened chaos in many departments of the social superstructure, which could only be put in order by international co-operation. The standardization of the clock around the world was indexed, by international agreement which records the British imperial supremacy of the epoch, to London and the Royal Observatory in Greenwich. It's a symbolic moment in modern history. You could almost say it represents a definitive break with the past, a change in the very measure of history. From now on, it will be possible not only vastly to improve the regulation of prices, but also to record exactly the moment at which events take place.

A delusion will grow: that it is a real gain in the quality of life to be able to take more effective control of time by measuring it ever more precisely. Such a delusion is fuelled by the growing precision of science. Soon, 'scientific management' will appear – the technique of timing the segments of the worker's labour process by which the output of his or her labour power may be more precisely measured. It is also highly relevant that this new regulation of time – and subsequently, similar international treaties concerned with the regulation of the flow of information in different media – transcends the barriers between nations, from *Realpolitik* to Cold War. The one thing opposing factions do not entirely break off are the channels of communication. The great problem is how to control them, how to prevent the enemy hearing what you don't want him to, how to monitor the enemy's communications, and how to influence his interpretation of what you let him hear. It is significant that the pioneering work in which Alan Turing was engaged during the war, which led to the construction of the first electronic computers at the end of the 40s, was the application of electronics to the mathematical art of code-breaking.

IV

Nor did the telegraph serve capitalist interests only passively. By linking distant markets together, the telegraph turned them into one vast interdependent market, in which a change in price in one part affected the whole system at once. As it advanced beyond the frontiers of the traditional markets, the telegraph helped to extend the geographical reach of capitalism. It also helped to intensify its operation. As a result of penny postage, railways, telegraphy, in short, the whole improved means of communication (as Marx observed in *Capital*), Britain already carried out five to six times more business with about the same circulation of bank notes. In short, new and improved means of communication were not just a range of products which entrepreneurial capitalism produced in its factories, but a necessary part of its social means of production. More accurately, they constituted an infrastructure, which stands to the individual producer as a precondition to his own undertaking, though at the same time this infrastructure is a product of the whole ensemble of individual enterprises.

Being an infrastructure, however, the development of communications constitute a problem for the development of the capitalist mode of production. They consist in *general conditions of production* which, as Marx explained in the *Grundrisse*, presuppose a stage of development of the forces of production and private capital for which their improvement is itself a necessary condition. Because of this, the matter becomes a special object of interest on the part of the state, in its role of overseer and arbiter of social development in the interests of the capitalist class; and thus it undertakes the necessary enterprises which at any given moment the capitalist class left to itself is unable to. This is already to be seen in Prussian and French government sponsorship of the first telegraph lines. It is a role, however, to which the state has to learn to adapt. When one of the inventors of telegraphy, a man by the name of Ronalds, quite naturally, in 1816, offered his services to the Admiralty, they declined. Half a century later, with anarchic competition between the different private telegraph networks leading to economic chaos, the British government decided to act, and empowered the Post Office to take over the entire telegraph system. It was the first nationalisation.

V

In 1944, the North American business magazine *Fortune* published an article on 'World Communications'. It warned that the future growth of the United States depended on the efficiency of US owned communications systems, just as Great Britain's had done in the past: 'Great Britain provides an unparalleled example of what a communications system means to a great nation standing athwart the globe . . .' The ideology of the growth of communications, as represented in the mass media themselves, holds them, like Alice, to have 'just growed'. As *Fortune's* warning demonstrates, they didn't.

Around the 1860s, the most progressive factions in the British ruling classes sensed how important it was to direct the growth of communications. Initially, the telegraph followed the spread of the railways. As modern industrialists themselves, the railway entrepreneurs employed the telegraph to improve their safety and control networks, also offering the service to private users. Soon, there were separate telegraph companies following the different routes of their big brothers, the individual railways companies.

However, there seems to have been a delay before it was generally understood that railways and the telegraph represent different types of infrastructure – the telegraph, though terrestrial, isn't a form of transport – and they were therefore destined to follow different patterns of capitalist development. The change in comprehension is recorded in legislation. During the 1860s, there were three Post Office Acts. The first, in 1863, merely defined the telegraph as a piece of wire. In the second, 1868, the Postmaster-General was empowered to 'acquire, maintain and work electric telegraphs'; the third, in 1869, amended the definition of the telegraph to include 'any apparatus for transmitting messages or other communications by means of electric signals'. This legislation, which effectively licensed a monopoly in an age which was deeply opposed to monopolies, was inspired by Gladstonian Liberalism – a political creed that was responsible for a series of expedient reforms in a number of fields, ranging from the disestablishment of the Church and the Education Act of 1870, to opening up higher civil service posts to examination and the 1872 Ballot Act.

If the case of the telegraph is a strikingly early example of nationalization, the results were not encouraging. There have always been awkward contradictions in a statist undertaking of this nature. As in almost every example of social-democratic nationalization ever since, the private companies were over-generously compensated to begin with, and the Post Office was unable to make the service run at a profit. A principal reason for this failure is the constitutional reluctance of the bourgeois state, having adopted nationalization, to follow the appropriate logic and properly take care of investment. With this very first nationalization, it also took time before the need for planning was understood. At the start it hardly seemed necessary: when it was nationalised, the telegraph system was still expanding, and for some time, all the Post Office had to do was go on opening telegraph offices all over the country (the number of telegrams sent in England and Wales grew from 7.1m in 1870 to ten times that number four decades later).

Incomprehension continued longer in the civil service than the Post Office itself, where new engineering concepts were gaining ground. However, the government's cost accountants, the Treasury, still thought in terms of the world of mechanical engineering, like bridge building and canal construction. When the telegraph had been invented, telegraphic messages, or telegrams, were at first regarded as apparently another form of mail, which is to say, a physical load, and this was still the spirit in which the 1869 Telegraph Act was drafted. When the telephone came along, and the Post Office wanted in, the definition needed to be stretched. They filed suit against the telephone companies under the terms of the 1869 Act, following the advice of government lawyers that telephone communications were telegrams within the meaning of the Act; and they proposed to the Treasury a plan for a comprehensive system of Post Office exchanges throughout the country. This the Treasury thought too expensive. The result was that while the GPO undertook a small number of local exchanges, more at first for business than for private use, much more important was the privately owned National Telephone Company, which grew by absorbing its other competitors. Not until 1912 did the Post Office take over the National

Telephone Company and acquire a near monopoly of the whole system. (Meanwhile, the GPO had already become the largest single employee of middle-class women. By 1911, 14,328 women were engaged as telephonists and telephone operators, 20,337 as counter assistants or clerks.)

The establishment of the Post Office telegraph monopoly can be related to the progress of the newspapers and of Reuters. According to one commentator,

> The key to Reuters' dominant market position in the sale of international news to the British newspaper press was its relationship with the Press Association, the national news agency established by the provincial daily newspapers and formally constituted in 1868 . . . The Press Association adopted the task of disseminating national news to its member-clients, and also lent its support to the campaign for the nationalisation of the telegraph, which came about in 1870 (Boyd-Barrett, p. 113).

But Reuters' carefully nurtured ideology of objectivity was to come under challenge from the emergence before the end of the century of the yellow press, trading on a radically different set of news values, that came from the world of commercial entertainment rather than the boardroom. The creation of the yellow press debased improved intelligence in the same way traditional popular culture was debased in the growing commercialisation during the last twenty years of the century of music hall entertainment, with the formation of syndicates as the first step in the development of entertainment capital, with its own sectoral interests: what the Frankfurt School people in the 30s called the culture industry.

A profound shift in social sensibilities was involved in this process, which involved major changes in the structure of the press that deeply affected Reuters. The modern press evolved in two main stages: first , to service the needs of the capitalist and professional classes for organs of information, and then, as instruments directed to those social classes and strata over which the capitalist classes needed to extend their domination. The creation of Reuters belongs to the former, that of the yellow press (which appeared in the 1890s) to the latter. As rival agencies grew up to supply the needs of the new mass readership papers in the period leading up to the First World War, Reuters suffered declining profitability, from which it had only partially recovered when the Depression brought fresh financial deterioration.

The 30s produced another worry too: authoritarian currents in government began to talk of the advantages of something more than the kind of informal arrangements between the news agency and the State that came into operation during the First World War. The Fleet Street owners, the Newspaper Proprietors Association, found themselves persuaded to take joint ownership of Reuters alongside the PA. Later they were joined by various Commonwealth Press Associations, and further ownership links were created when Reuters became part owner together with the BBC and the television corporations of the white Commonwealth countries, of the world's foremost television news agency, Visnews.

VI

The growing interaction of the scientific, the technical (or technological) and the financial (or economic) is clearly demonstrated in the way in which science was beginning in the 1860s to be integrated into the economic system through a number of new companies, the first purely scientific commercial enterprises, which manufactured the equipment for telegraph, cable and the telephone. The development of these companies had a chain of social effects. They created new professions, such as the electrical engineer. They provided what Bernal called 'the stock-in-trade for electrical experimentation – batteries, terminals, insulated wire (Faraday had to use wire from milliners or wind his own insulation), coils, switches, simple measuring instruments – and all at prices which even impoverished university laboratories could afford' (Bernal, p. 117). Soon these enterprises set up their own research departments, inventing a new business practice nowadays called R&D. The telegraph led directly to the telephone, and later to the wireless telegraph. It provided (Bernal again) 'a nursery for the young science of electromagnetism, supplying problems, part-time occupation, equipment and funds for the academic scientists and ensuring them plenty of students' (Bernal, p. 23). The telegraph and cable industries were also the sources of the new electric light, traction and power industries of the 1880s and 90s.

Bernal also makes the observation that the telegraph led to the addition of new electrical units of measurement to the age-old weights and measurements of trade and commerce. This is crucial. The telegraph was the first technology to establish information as a commodity, and therefore the need to measure it. Conceptually, this is the first step towards the modern distinction between hardware and software, and it belongs to the analysis of the peculiarities of the commodity form of all the electrical media, and later electronic media: including the telephone, phonographic recording, film, radio and television, which all got sucked into the same process. With each different medium you get new variants on the basic facts of technical linkage: if you've got something you can call software – the information you're passing around – then you've got to have something called hardware – the medium you pass it through.

This is reflected in the phenomenon of commodity linkage, which takes different forms, like cameras and film, or gramophones and records. Since you can't have one without the other, this gives rise to a general principle, namely, that manufacturers of any new kind of hardware have to concern themselves with the production of the appropriate software without which the hardware has no market. In this way, early producers of cinematograph equipment were also film producers and distributors – the very distinctions took time to appear. Or the early recording companies made both phonograph and phonogram; some still do. This can also be compared with the relationship between broadcasting and programmes. The first manufacturers of radios had set up radio stations *and* produced programmes, as a kind of loss leader, until the means were found to relieve them of the need, and institutional or commercial broadcasting began.

Observe that software becomes an ambiguous term here, on the one hand referring to the content which passes through the medium, on the other to the physical form which the content takes, the record on (or in) which the content is contained (or the modulated radio wave). Or it may not refer to the content at all. In the case of photography, the camera is no use without film, but you take your own pictures.

The ambiguity of the term software isn't just loose thinking; it comes from the fluid and shifting relationships between form and content which are characteristic of the media. Because the content of communication is symbolic, it is possible to translate it between forms in various ways, so that the different media become devoted to preying off each other: newspapers consume photographs, radio consumes records. In fact the relationship between the latter pair is thoroughly symbiotic. Radio needs records to help fill up its air space, but the recording industry uses radio as an aural shopwindow to publicize and plug its ware.

These patterns relate to another series of peculiarities, in the commodity nature of the media, which have to do with the various different ways of consuming cultural products and therefore the different kinds of exchange value which are yielded. Thus, while cinema imitated the performing arts in collecting gate money – the cash paid in at the ticket window – and gramophone records imitated books, there were no pre-existing equivalent for broadcasting. The first manufacturers of radio not only had to produce programmes, they had to produce them gratis, because there was no way of selling them. Indeed, with the exception of pay-tv, programmes are still not exactly commodities from the point of view of the listener or viewer, but more like a right which comes with the purchase of the set, and in some countries, the payment of a licence fee. To finance broadcasting through a licence fee is one solution. Commercial broadcasting is another, which in the process creates another new kind of commodity: the airspace which is sold to sponsors and advertisers. It has even been said that the real commodity isn't the airspace, but people, the audience, according to the statistical breakdown of the consumer polls. [. . .]

[T]here are special kinds of property rights involved in this whole process. Copyright is a concept which has constantly shifted its meaning ever since it was first defined, in the wake of the invention of printing, to answer the question of the ownership of a text in terms of who held the right to make printed copies of written works – which at the beginning meant the printer-publisher, not the author. The history of copyright is the history of mounting contradictions in the legal superstructure, as the changing forms of cultural production altered the social relations of the author, who gradually became a new kind of intellectual worker (a shift which relatively few of them recognised until after Schiller and Marx). From the intellectual worker there has now descended the alienated mental worker whom Schiller and Marx foresaw; from the critical consciousness of the artist and the scientist, there have descended the programme producer and the computer programmer. Freelance programmers of all kinds, of course, still have an awkward tendency to organise themselves to claim the ancient privilege of copyright.

VII

This growing infrastructure has contributed independently to social relations by creating new sensibilities, new ways of relating to the world, and of representing it. Information is the lubricant of the capitalist mode of production, but at the same time it creates its own symbolic domain. If the human being is, as Merleau-Ponty said, condemned to meaning, then our mental existence cannot but reshape itself around the new languages and dialects which now occupy our world. We should think of this too as part of the Reuters Factor. It isn't just a question of information generating business, but of the nature of the information business itself, the way that gathering information treats the world.

For instance, the growing demand for information has created new metiers. Harold Perkin has suggested that statistics, which emerged as a discipline during the industrial revolution, 'is to industrialism what written language was to early civilization: at once its product and its means of self-expression' (Perkin, p. 326). The Utilitarians, who promoted the Statistical Societies which exploded into activity in the 1830s, saw the role of statistics as, according to Perkin, the 'discovery and examination of "intolerable" facts, often long before they were felt to be intolerable by the press and public opinion'. But statistics also introduced the practice of surveillance, both commercial and bureaucratic. The kinds of information involved in this process come to impose their own terms of reference, which in turn become one of the ways in which capitalism represents itself to itself, seeming to impose order and reason where there is none.

VIII

The strategies adopted by the newspapers relate to their dual character, as organs of information on the one hand, and of influence on the other: their character, in other words, as synthetic forms of cultural production. This is entirely typical of the modern media, one might say symptomatic. It certainly helps to explain how social susceptibilities are shaped in ways that people aren't normally consciously aware of. The social unconscious is formed of many elements, reaching back through the diverse heritage of popular culture as well as the impact on traditional belief of the successive shocks of social 'progress'. The influences which shaped the character of the mass press in Britain include, for example, such features of popular culture as the broadsheet ballads and the art of the patterer, which had both served for the dissemination of news.

The traditions of popular culture were sustained through the development of new forms of popular entertainment, like music hall. To begin with, musical entertainment was subjected to progressive commercialization without any help from new communications technologies until almost the end of the century. Then, the phonograph, which first appeared in 1877, introduced another peculiarity: it was primarily a cultural phenomenon, which did little to increase the circulation of information – on account of which, it was much slower to develop into a major branch of capital. Its connection with the complex of modern scientific-based industry is amply shown by its birth in Edison's research laboratory, but for several decades it seemed to lead a more or less independent existence as a minor branch of the entertainments industry, and was only recapitalized and reconverted with the development of electrical recording in the 1920s, after the invention of electronic amplification during the First World War.

The phonograph brought about a cultural revolution in the interim nonetheless. At the beginning, after the invention of the first mechanical means of sound recording by Edison in 1877, economic exploitation was impeded simply because there was no mechanism for duplication of the recording. The early cylinder machines made good side-shows in the fairgrounds, and they had the attraction for the private purchaser that you could make and show off recordings in the home, but this hardly provided a mass market – much more restricted, for example, than the telephone, which was also sold, in its early years, as a luxury item for the home.

But the phonograph added nothing to the communications apparatus, because it didn't carry information; its improvement was therefore less urgent, and didn't attract the same investment funds as other new instruments. That is why its massification had to wait thirty years. The industry only took off after Emile Berliner accomplished a series of improvements during the course of the 1890s, culminating in the first wax disc recording in 1900. The wax disc served as a master for a copper matrix from which copies could be made. Now there opened up a new and enormous market. Its nature can be judged from a trade advertisement put out by the New York-based Victor Company in 1905, with photographs of leading recording artists and a text which explained: 'Three show pictures of operatic artists, one shows pictures of popular artists. Three to one – our business is just the other way, and more, too; *but there is good advertising in grand opera*.' The gramophone now began to do for music what telegraphy had done for information: it created new conventional forms, both standardised and truncated, extended the reach of the market, and increased circulation.

Again the Reuters Factor was at work. The record found a much larger audience than the artiste could reach in person in theatre or concert hall, both in terms of numbers and of geographical extent, and before long the record industry took on an international character. Music has always travelled – people carry it with them – but now began a process of wholesale trade which transplanted music from one place to another whatever the cultural predilections and differences. It also overrode the musical cultures it penetrated by imposing its own increasingly industrial nature, which Adorno in particular has analyzed, seeing it as a process of fetishization of musical characteristics in a way that negates their aesthetic authenticity. The result has been to transform the social role and functions of music. In 1967 a Mr Joseph Klapper of CBS was able to tell a US Congressional Committee, inquiring into 'Modern Communications and Foreign Policy', that 'the broadcasting of popular music is not likely to have any immediate effect on the audience's political attitude, but this kind of communication nevertheless provides a sort of entryway of Western ideas and Western concepts, even though these concepts may not be explicitly and completely stated at any one particular moment in the communication' (Schiller, p. 106).

IX

For all this to happen, however, the gramophone needed the radio. The case of Guglielmo Marconi and the invention of radio is [an] example of how the interest of the military in improving their communication systems comes to be of crucial importance at the earliest stages of development of a new communications technology. Marconi, like similar pioneers, had been able to achieve primitive wireless transmission over short distances as a very young man working at home with his father's resources, by 1895. But then he needed financial support on a considerable scale. When the Italian navy turned him down, he found the support he needed in England, where his mother came from, and his family connections were the right ones. He rapidly had the Post Office, the War Office and the Admiralty all involved in the development of the invention.

Important early steps for commercial wireless telegraphy included the first ship-reporting service which Marconi supplied for Lloyd's of London in 1898, and in the same year, the first journalistic use, when the Dublin *Daily Express* decided upon a publicity scheme and used the wireless to report the Kingstown Regatta. Two years earlier, when Marconi arrived in London, was a key year in the development of mass communications and mass entertainment in England, the year of the first cinematograph shows and the birth of Alfred Harmsworth's new *Daily Mail*. Marconi made skilful use of newspapers as a medium of publicity in promoting his invention on both sides of the Atlantic, and they were eager enough to be used. New inventions were a great source of public wonder, and even better, of circulation. The latest scientific wonders were showmen's acts. The man the English claim as the inventor of cinematography, William Friese-Greene, who didn't do too well with his invention, was reduced to treading the boards demonstrating the incredible new x-rays. This is extremely ironic when you consider that music hall was the perfect launchpad for the film – together with the fairground and the tradition of the travelling showman – and that once films took off, they slowly strangled their host.

It was the cinema, more than any previous invention of new communications technology, which suddenly changed the ground-rules. Film was from the very beginning peculiar in two things: it established itself first with the mass audience, and only then filtered upwards through society. This can be compared with the immediately preceding inventions of the telephone and the phonograph. Both of them found their first markets within the bourgeoisie and only later became working class commodities. You can see this in the earliest publicity announcements for the telephone, which spoke of business needs first, of its social usefulness second, and of its value as a personal luxury item third. In the case of the gramophone, as Berliner renamed it, its initial appeal was also found among social groups with special cultural interests, since even after discs first appeared, the period of primitive pre-electrical technology in the industry was protracted: with the lack of improvement, the original spontaneous fascination of the populace slackened, and it helped to be able to keep a market of immigrants in New York happy with recordings of Italian opera.

But film had no such problems. Its appeal was not only immediate and huge, but all such cultural differences were submerged in it. The fact that its characteristics were primarily aesthetic, and it added to the apparatus of commercial intelligence no more than the gramophone, hardly mattered in the face of its rate of growth. Because of this, and because no country in which film appeared was capable of producing enough to supply the market, film was also international from the outset; by the time of the First World War, it had begun to attract the interest of finance capital. It was the equivalent in the sphere of cultural production to the transnational character of the electricity industry, which Lenin held to be ushering in imperialism – the highest stage of capitalism.

X

We have surveyed the nineteenth century and what do we find? First, that the growth of industrial capitalism, and indeed, of imperialism, is intimately linked with the invention and development of new means of communication. It is a process in which military interests play a significant role, though not always as central as in the twentieth century, the epoch of neo-colonialism. The needs of commercial intelligence, however – in other words, of the bourgeois class itself – are constantly very much to the fore and provide the initial markets. Here is a parallel with the Thatcherite approach: Thatcherism looks to the application of the new high-tech in the form most beneficial to corporate capitalism itself. A rationale results: to achieve this application, corporate capitalism must be free to do whatever it wants to.

In the nineteenth century, the new communications begin to form a new infrastructure, which stands, however, in contradiction with the ideology of entrepreneurial capitalism, principally because it needs state regulation and control if the piecemeal initiatives of individual entrepreneurs are to be effectively welded together. Hence we find the first steps in international cooperation designed to establish certain basic universal standards, but in each new branch of communications there is a bitter fight between capitalist competitors before the victors are able to establish technical standardization. This is, of course, extremely wasteful. Yet in the Thatcherite vision of the coming of high-tech, there is to be the same chaos in the marketplace, the same destructive wastefulness.

We also saw that this new infrastructure progressively opened up new economic opportunities for the exploitation of cultural production. For example, alongside the commercialization of popular entertainment like the music hall, and the creation of the first mass readership markets, the 1880s saw the hugely successful introduction of the Kodak camera – 'You press the button, we do the rest', which made photography the first new popular art form of modern times. As the century draws to a close, the pace of innovation accelerates. Photography waited forty years for its massification. The delay in the case of the phonograph is less than thirty; and by this time, cinematography has burst upon the world. In the course of these developments we find a distinction growing up between communications technologies where the primary function is the transmission of intelligence which has a practical or instrumental use value, and those where the content is primarily symbolic – which are primarily, in other words, new means and media of cultural production and reproduction. Both, however, operate in certain respects in curiously similar ways. In both cases, there is a necessary distinction between what is later called hardware and software, though to be sure, each technology has its own peculiarities (just as the products of different media have different peculiarities as commodities, in the way they realize their exchange value).

The development of electronics during and after the First World War radically modifies the separation between different media and reshapes the entire communications industry. With radio, talkies and television, there emerges a culture industry in which all branches of communication are implicated, and in which intelligence and information plays second fiddle to the universal levelling of mass consumption. Their function – most starkly seen in the United States – is not to produce an informed and educated population, with a higher cultural level, but to shape their consciousness to the needs of increasing passive consumption, in which the products of the media themselves consume a greater and greater proportion of consumer spending. Thus the economic development of the media – involving, as always, the multiplication effect of the Reuters Factor – extends and intensifies still further both the information network and their own cultural net. To this is now being added the impact of microelectronics, which is profoundly contradictory.

The progressive enhancement of potential cultural production brings inevitable new contradictions. As Hans Magnus Enzensberger observed fifteen years ago, it is wrong to regard the media merely as means of mass consumption. They are always, in principle, also means of production. The contradiction between producers and consumers in the mass media is not inherent but institutional, and it has constantly had to be reinforced by economic and administrative measures (including the appropriate design of the equipment itself). But the original massification of photography was precisely a matter of placing a new cheap means of cultural production into the hands of the masses. Such an enterprise can be very risky; it had been necessary to repress a radical press that appeared during the industrial revolution, and to seek to control the self-education movements which the new industrial proletariat had created. Yet although photography, then 16 and 8mm cine, and now computers, are accompanied by their own sub-industries devoted to the culture of the amateur, there have repeatedly been eager users who fall beyond the pale, and create alternative uses and networks of users. It is to the defence of alternatives at every level that the critique of Thatcherism must be directed.

Many commentators have observed that it is a misnomer to speak of the communications media: the media are used to prevent communication. But this is precisely because what Brecht said of radio in 1932 not only remains true, but in the age of microelectronics and the computer, and the promise of 'interactive information/entertainment', becomes even more pertinent:

> Radio must be changed from a means of distribution to a means of communication. Radio would be the most wonderful means of communication imaginable in public life, a huge linked system – that is to say, it would be such if it were capable not only of transmitting but of receiving, of allowing the listener not only to hear but to speak, and did not isolate the listener but brought about contact. Unrealizable in this social system, realizable in another, these proposals, which are, after all, only the natural consequences of technical development, help towards the propagation and shaping of that *other* system . . . If you should think this is Utopian, then I would ask you to consider why it is Utopian.
> Brecht, *Theory of Radio* (1932).

References

J.D. Bernal, *Science and Industry in the Nineteenth Century*, Routledge & Kegan Paul, 1970.

Oliver Boyd-Barrett, *The International News Agencies*, Constable, 1980.

Hans Magnus Enzensberger, 'Constituents of a Theory of the Media', in *Raids and Reconstructions*, Pluto, 1976.

Karl Marx, *Capital* I. Lawrence & Wishart, 1970.

Harold Perkin, *The Origins of Modern British Society 1780–1880*, Routledge & Kegan Paul, 1972.

Herbert Schiller, *Mass Communication and American Empire*, Boston, Beacon Press, 1971.
Graham Storey, *Reuters' Century*, Max Parrish, 1951.
'World Communications', *Fortune*, May 1944.

3.2 Claude S. Fischer, *The Revolution in Rural Telephony 1900–1920*, 1987

It is a common intellectual habit to amalgamate many of the daily implements of contemporary life – cars, video recorders, refrigerators, assembly lines, and so forth – into a single class, labelling it 'modern technology,' '*technique*,' 'rationalization,' or the like, and then crediting that class of objects with helping bring on 'modern' culture or personality. This jumbling together rests on some dubious theoretical assumptions and on the absence of empirical research about specific devices. The telephone is one such specific technology, typically listed, in an undifferentiated way, among other 'modern' objects, and, also typically, unstudied.[1]

A close examination of the telephone's history – and presumably of any particular technology – suggests that technological change is more complex. Rather than being part of an inevitable unfolding of modernization, the development of telephony was contingent and discontinuous. Rather than being part of a comprehensive rationalization imposed upon its users, the telephone was sought or rejected, adopted or abandoned, and employed in various fashions according to the interests and perceptions of specific consumers. Or so suggests the history of rural telephony in America. The telephone's role 'belongs to the important and subtle class of problems in the social sciences which demands a logic more complex than simple causality' – i.e., that technological development impels its use and 'impacts' people – but instead 'a logic that allows for purposive behaviour as an element in the analysis.'[2]

In the case of rural telephony, we see a diffusion that proceeded in fits and starts, a process highly dependent on legal, business, and political circumstances often having little to do directly with the technology itself. We see a startling reversal of the commonplace that modern technologies spread from the advanced sectors to the backward ones. Whereas most 'modern' innovations are last adopted in rural places,[3] for a time, more American farm families had telephones than did town families: In 1920, 39 percent versus about 33 percent or less. And in this case study we see the role of human agency: people seeking out a technology, independent of or even in opposition to its corporate sponsors, and adapting the technology to their own needs.

After briefly reviewing the general history of telephony in America, this article will describe the rise of telephony on American farms and the relationship of the telephone industry to farmers; analyze the reasons why farmers adopted the telephone so enthusiastically; and then compare this case to that of the automobile.

Source: *Journal of Social History*, vol. 21, 1987. pp. 5–26.

The historical context[4]

Gardiner Hubbard, Alexander Graham Bell's father-in-law and major backer, made two fateful decisions soon after he began marketing Bell's telephone in 1876: that telephones would be leased rather than sold and that the Bell Telephone Company would franchise its monopoly service, taking fees and, later, stock equity in return. Leasing (ended only recently) kept control over Bell's basic resource, the instrument – patent infringers were vigorously prosecuted – and helped integrate all the vendors. Franchising (ended in 1984) guaranteed that the dominant telephone companies throughout the nation would be centrally controlled.

Both decisions also meant that diffusion would initially depend on the initiatives of local telephone entrepreneurs. Those entrepreneurs approached businessmen, by and large, and their subscriber lists grew rapidly at first: the number of telephones tripled between 1880 and 1884. But telephone development in the United States leveled off during the next several years, less than doubling between 1884 and 1893, reaching 266,000.[5] During this period, the Bell Company, under the general management of Theodore N. Vail, improved the technical performance of the system and developed the first long-distance, or toll, lines. Vail also strengthened control over the franchised companies, so as to insure Bell's hold on the telephone business beyond the life of the patents, and in other ways centralized the industry at Bell headquarters.

The original telephone patents expired in 1893–94 and thousands of new telephone operations emerged in less than a decade. Bell's competitors were of three types: major commercial companies; small 'mutual companies' set up by users to serve themselves; and 'farmer lines', simple cooperative systems, often built without regular switchboards. The Census Bureau estimated that in 1902 there were about 3,000 of the first, 1,000 of the second, and 5,000 of the third (75 percent of these were in just four states).[6] Some of the farmer lines literally used fence wire to pass the current from farm to farm. Most of the new operations were undercapitalized, hurriedly built of poor materials, and evanescent. Typically, they served areas that Bell had ignored.

Nevertheless, the new enterprises, the occasional head-to-head competition with Bell, and the price-cutting they brought on spurred rapid diffusion: almost a nine-fold increase in telephones per capita between 1893 and 1902, as compared to less than a two-fold increase in the prior nine years.[7]

Until 1907, Bell's response to this challenge was to compete fiercely at all levels: It cut its charges to 'fighting rates,' at some places below cost; refused to sell equipment to, or connect with, other companies; decried the fiscal integrity of its competitors; argued in public forums that 'dual systems' in the same territory were wasteful; and otherwise did whatever was needed to control franchises and drive out competitors. Bell also innovated ways of attracting new, usually poorer, classes of subscribers: party-lines, coin-box telephones, and further extension of 'measured service' (charging by the call, which lowered annual rates for most users).

The competition, the capital requirements of expansion, and the increasing debt of a company rapidly losing its market share (from

100 percent in 1893 to 49 percent in 1907) led to various reorganizations of the Bell system. In 1907, J.P. Morgan and his New York allies gained control and reinstalled Theodore Vail, who had left the industry for many years, back at the top of AT&T. Vail changed strategies. Rather than recklessly expanding into new territories and cutting prices to forestall competition, AT&T bought out competitors where it could and ceded areas where it was losing. Publicly, Vail argued that the telephone was a 'natural monopoly' and that there should be 'universal service' – which AT&T was best-placed to provide. In return, he acceded to regulation and worked to correct AT&T's image as a callous trust.

The 'independents' were stymied. Although often competitive in cost and quality with Bell, they had difficulty expanding beyond their small-town bases, could not establish their own long-distance lines, and were especially hurt by lacking connections to New York City. Many were guilty of poor financing, poor planning, and poor service; others accepted or even solicited buy-out offers from AT&T or its allies. Between 1907 and 1912, the Bell System had reversed trends and regained an additional six percent of the market.

At the same time, state regulation, including the requirements that all telephone systems interconnect, was increasing. In an apparent effort to forestall a Federal anti-trust suit, AT&T issued the 'Kingsbury Commitment' in December, 1913. It agreed to divest itself of the recently purchased Western Union, to cease acquiring companies directly competing with its own, and to provide long-distance connections for non-Bell systems. Over several years, local telephone service was divided into geographic monopolies. Where direct competition had occurred, swaps of company plants were arranged. (For example, in 1917 Bell traded its plant in Rochester for the independent system in Buffalo, New York.) By the mid-1920s, industry-wide cooperation had reached the point that leaders of the independents described their earlier competition with the Bell 'octopus' as an unfortunate error.[8]

The modern American telephone system – predominantly Bell local service and exclusively Bell long-distance service – was essentially fixed from the 1920s to 1984. This system was consolidated, technically improved, and by 1929, 42 percent of all households had telephones. The number of telephones in service shrank during the Depression, by over 3 million, from 41 percent of households to 31 percent. But from 1933 on, telephony in America expanded steadily, reaching 37 percent of all households in 1940 and over 90 percent in the 1960s.

The story of rural telephony[9]

The first telephones were set up in cities. Their vendors hardly notified American farmers, much less made any effort to sell telephones to them (mocking anecdotes of rustic incredulity over the talking machine were common in industry journals). Bell concentrated on the major urban centers. But after the novelty wore off in the countryside, farmers and villagers in many places requested service from Bell licensees. In Centralia, California, for example, businessmen petitioned the local franchise for service in 1891. In 1892, in Chico, California – as happened elsewhere – an unauthorized entrepreneur set up a fence-wire system for unserved local farmers unable to get service from the Bell subsidiary. We have no way to judge the scope of unsatisfied rural demand for telephone service, except by what happened after Bell lost its patent monopoly.

When the patents expired in 1893–94, rural telephony grew rapidly and did so from local roots. By the first full telephone census in 1902, there were at least *6,000* tiny farmer lines and cooperative mutual systems. Their telephones, plus those of commercial companies operating in rural places, numbered at least 125,000, or 5.3 percent of the national total, at a time when 59 percent of Americans were rural. Five years (and a moderately better count) later, there were about *18,000* independent farmer lines and mutual systems, and an estimated 1.5 million rural telephones, 24 percent of the total. An estimate for 1912 claimed that 3 million, or 38 percent, of all American telephones were rural, at a time when 53 percent of Americans lived in rural places.[10]

The best comparative estimates of farm telephony are that, in 1902, *two or three* percent of farms had telephones, compared to about ten percent of all American households. In 1912, roughly *30* percent of farms had telephones, as against *31* percent of all homes. And in 1920, *39* percent of farms had telephones, versus *35* percent of all American households. Regions varied greatly in levels of farm telephony, with farmers in the Plains States reporting subscription rates of over 80 percent and those in the South of less than five percent in 1920. (And Canadian farmers never approached the levels of subscription attained by urban Canadians.)[11]

Thus, US farmers, virtually without any telephones in 1900, had reached a diffusion level of almost 40 percent in 20 years. More striking, they had rapidly surpassed the level of city-dwellers during that period. This performance is yet more impressive given the disadvantages farmers faced, especially their much lower incomes and, as we shall see, the indifference of the major telephone companies.

Typically, a rural mutual system or farmer line was organized by a group of leading farmers, or a small-town merchant or doctor, whose efforts to solicit service from a major commercial company had failed. For an initial investment of $15 to $50 and often their own time and materials, roughly 15 to 50 farmers would combine as shareholders in a mutual stock company, receive a telephone, and connection to others in the rural neighborhood. Annual rental fees might run from $3 to $18 a year, less if the subscriber was a shareholder. (Shareholders, however, were often assessed for needed capital.) If the system had a switchboard, a farm wife or daughter typically served as operator during the daytime. Often, the shareholders arranged a connection to a commercial, or larger mutual, company's switchboard in town, and through that, to the wider world.

Most such systems remained small, under 25 or so subscribers, although some grew larger or were consolidated: Most eventually served as 'feeders' to larger systems, including Bell's. Typically, these operations survived on a small cash margin, with marginal equipment, and postponed maintenance. Service was frequently poor, interrupted, or lost for long periods. The amateur business managers usually underestimated costs, particularly of depreciation, and therefore undercharged for the service. The systems also often depended on subscribers maintaining their own telephones and incoming lines, which many farmers failed to do. Despite all this, the number of such systems and their subscription lists grew enormously.

Relations with Bell and large independents

The Bell companies largely ignored rural and small-town America before the era of competition. Eventually, AT&T did attend to farmers, but largely did so defensively and reluctantly. In order to protect its urban, commercial customers from the seductions of its competitors, Bell began, at about the turn of the century, to serve farmers, even if occasionally at a loss: The company that had farm subscribers thereby had strong leverage in signing up farmers' suppliers and customers.[12] In some places – notably, New England – the Bell System responded to the surge of rural systems with 'Rural Service' plans. The local Bell company would help farmers build lines and would provide them with exchange service. With these plans, rates for farmers dropped from the prohibitively expensive charges of the pre-1900s, a minimum of $50 a year in 1893, to $10 a year in 1903.[13] (In non-Bell areas, rates were usually lower.)

Internal AT&T memoranda clearly explain that farmer service was seen essentially as a competitive strategy in the 1900s. For example, a report from Pacific Telephone and Telegraph in 1908 provided this account:

> Prior to the year 1905 this company made no special effort to encourage farmers to build lines to connect with the Company's exchanges but endeavoured to meet the demand by building and maintaining suburban lines [presumably with high mileage charges[14]], and furnishing exchange service at rates varying from $1.50 to $2.50 per month.
>
> . . . In the early part of the year the fact developed that opposition could successfully secure our business in small exchanges by first canvassing the adjacent rural territory and, by inducing the farmers to join with them, force the town people to subscribe also.
>
> To combat this opposition the Company established a fighting rate of $1.00 per year, covering exchange service and the loan of an American Bell transmitter and receiver . . . [A]t several exchanges where we have a numerous farmer subscription . . . opposition has been unable to enter . . . while . . . where we had but few farmers connected, our exchange was practically wiped out by the opposition.
>
> Aside from the benefit of protection, there can be no doubt that, at the lower rates, we are furnishing service below cost.

The report claims a gain of over 300 percent in farmer subscriptions between 1905 and 1908.[15] Reports such as these show a continuing reluctance to provide rural telephones – at least at the rates that comparison-shopping farmers were willing to pay – without the spur of competition.

Bell's campaign for farmers changed with the ascendance of Vail. He conceded currently undeveloped rural territory to the independents. And he authorized the first connections between Bell exchanges and non-Western Electric equipment. This accelerated efforts begun in some places to encourage farmers to build their own lines for connection to the Bell System – thereby both capturing potential clients of their competitors and saving on construction costs.

As competition faded after the 1913 Kingsbury Commitment, the impetus for rural service seemed to come largely from government regulators. It still remained somewhat unappetizing to Bell companies.

In 1916, J.D. Ellsworth, AT&T's Vice-President for advertising, convened an afternoon session of a Bell System conference with the comment: 'I believe that there is good reason for specializing on this particular class of people [rural residents]. They are numerous, they are indispensable and they are politically powerful.' He asked the conferees – advertising managers of local companies – to discuss what they were doing to woo farmers. With rare exception, one after another answered that little was being done to generate farmer business, that their companies simply responded to guaranteed applications, and that rural advertising, if any, was directed to creating an appreciation of the true cost of the service – i.e., explaining their rates. Ellsworth concluded: 'Mr. Vail has frequently boasted of our high rural development and I do not want him to be boasting without grounds . . . In spite of the high cost of material, we must not overlook the farmers.'[16] Indeed, institutional advertisements of the period bragged that farm telephony demonstrated AT&T's commitment to 'universal service.'[17]

But AT&T and its subsidiaries did not seem to try hard to sell telephones to farmers, except for the rate-cutting and the organizational work undertaken during the era of fierce competition. At times, suggestions and efforts were made to dispose of lines serving farmers. And much effort was spent pressing for higher rates to farmers. Into the late 1930s, farmers in predominantly Bell territories were less likely to have telephones than those in areas where Bell was not dominant.

The telephone companies outside the Bell System were more interested in, but still ambivalent about, farmers. Small-town and rural America was the natural market for the independents, because it had been unserved, or imperiously served, by the Bell monopoly. Still, the profits seemed to be in the towns, so farmers were often considered a necessary burden – key to producing town business, yet difficult to deal with. One independent summed up the general strategy in 1908:

> It stands as an undisputed fact that the development of the rural telephone has been the most potent weapon in the hands of the independents in their fight, first for an existence, and then for the successful fight for supremacy in the field of telephony.

He attributed Bell's 'defeat' to their 'constant refusal' to recognize rural telephone demand.[18]

Yet the independents, too, had reservations. Common skepticism about farmers' interest in telephones was expressed by the president of one company: '[T]he farmer was so unfamiliar with the telephone as to be absolutely frightened with [*sic*] it.' This notion was widespread even among those in the independent companies, such as this man, who most relied on rural customers. In 1902, an independent telephone journal reported that 'the farmers of the country were the very last class of people thought of as possible users' and one of its contributors claimed that, until recently, 'farmers practically did not know what a telephone was.' In 1909, independent commentators looked back on the 'early days' and wrote that farmers had then viewed the telephone only as an entertainment. In general, the independents' internal literature described farmers as ignorant, hard-headed, short-sighted, and tight-fisted – poor customers. The ambivalence is captured by one telephone businessman's words of encouragement to his colleagues: 'Do not let anyone convince you that you can not reason with the Iowa or Nebraska farmer . . . Do not be afraid of them!'[19]

Competing with Bell for connections to these mutual lines became a major headache for the independents. One Michigan telephone man, for example, praised individual rural subscribers, but complained that the mutual companies get 'the big head.' 'They think that because they have one hundred to two hundred members, or even a less number, that . . . they can dictate any terms.' Nevertheless, '[we must] work out some plan whereby we shall be able to keep our farmer friends with us and not let them be led astray by our enemy [Bell] . . .'[20] The independents frequently saw farmers, whether collectively or individually, as difficult partners. One independent man's paraphrase of farmers' demands conveys the general frustration: '[The farmer] wants free exchange with Afghanistan, Belochistan, and other foreign countries. Did not agree to pay anybody any toll – the phone doesn't work right anyway – think we'll go over on the other line – can get Ottumwa for a nickel there and you are a robber anyhow.'[21]

Rural telephony involved many unique or exacerbated problems – at least in the view of the commercial companies. The pages of *Telephony*, the major journal of the independents, and to a lesser extent, the internal reports of AT&T, are laden with complaints: that farmers 'kicked' about paying fair rates, avoided bill collectors, built inferior connecting lines, cared little about maintenance and repair, knew nothing about financial affairs, demanded rates for using high-quality toll lines as low as those they paid for their own jerry-rigged systems, listened in on each others' calls (thereby weakening the signal and starting fights), hogged the party lines gossiping or playing musical instruments, and were generally a rude and bumptious lot.[22]

The manager of a Michigan independent company wrote his rural subscribers a letter in 1908, perhaps capturing the range of aggravations his colleagues felt:

> Dear Subscriber:
>
> . . . [L]istening to other people's conversations over the telephone is as ungentlemanly or unladylike as it would be to creep up to other people's doors . . . and listen to a private conversation . . .
>
> . . . [C]onversations should be limited to five minutes, for, as you know, oftentimes you want to use the telephone when it is in use by others . . .
>
> Musical instruments should not be used to entertain over rural or party lines . . .
>
> It is true when you wish to use the telephone you should place the receiver to your ear to make sure the line is not in use . . ., but it is not necessary to stay there more than an instant . . .

> Some people take advantage of the telephone to say things to others they would not think of saying were they face to face with them. This shows a cowardly spirit . . .[23]

Although the independents also responded to, rather than initiated, farmer demand (except in places of fierce competition), they sought the farmer's business more than did Bell. Yet this, too, seems a reluctant courtship.

Given the actual demand and the eventual level of diffusion, why did the telephone vendors attend so relatively little to this market? The simplest and commonest answer is that there was no profit in rural telephony. AT&T President Gifford said in 1944 that it was difficult to provide rural service at a price that customer felt he could afford. The Publicity Manager for Northwestern Bell said more bluntly about 30 years earlier that '[t]he average farmer wants service at less than its costs . . .' His colleagues stressed repeatedly the high cost of providing service to farmers.[24]

As to those costs, independents and Bell alike claimed that it was considerably more expensive to serve – or serve well – rural users than urban ones. There is debate on this point. Some costs are higher in rural areas, such as poles and trunk lines. On the other hand, salary costs, land rent, construction, and other costs are lower. A 1984 Congressional Budget Office study concluded that telephone service did not clearly cost any more to provide in rural than in urban areas. Another critique of the industry position is that calculations of costs and revenues are meaningful only at the entire system level and therefore, in this case, part of the profit from urban business telephones should be attributed to their access to rural subscribers. As early as 1903 an observer noted that farmer lines increased the value of town lines. All this aside, it remains that rural costs were commonly perceived as significantly greater than urban ones, even by state utility commissions that set seemingly subsidized rural rates.

Such issues of profitability may also explain telephone men's general attitudes towards farmers. Farmers often demanded free toll service from their own switchboards to the nearest town, undercutting much of the commercial companies' profit base. The farmer lines' technical deficiencies posed problems for the larger systems; poor connections to rural subscribers were often blamed on the bigger companies. And accommodating the technical level of farmer lines – or the level farmers would pay for – may have undercut the high-quality service the larger companies (especially, Bell) sold to their business customers.

The rural market also probably offered poor opportunities for marketing high-profit items like extensions, PBX systems, and long distance calls.

Others contended that urban demand fully absorbed the capital available for new construction; none was left for rural areas.[25] No doubt true at times, this alone would not explain why comparable rural demand was ignored even in slack periods.

More fundamentally, many in the industry claimed that farmer demand, even for basic service, was limited and inelastic; rural people did not want the telephone enough to pay its true cost. How was this view reconciled with the surge of demand in the early 1900s and the industry's advertisements claiming that the telephone was a great boon to farmers? In various ways. A common theme was that the farmer responded only to short-term monetary calculation. For a low-income group usually short on cash, spending was no doubt severely rationed. Yet, poor farmers were willing to pay more of their income for telephone service than were comparable townsfolk. And independent studies of farmer demand do not point to rates per se as a barrier to subscription, so much as rates for poor quality service.

The argument about what was the farmers' 'real' demand reached an ironic state in the 1940s, when telephone companies resisted Federal legislation for REA loans to telephone cooperatives. Industry critics and farm representatives argued that rural Americans desperately wanted telephone service (at reasonable rates) and telephone company representatives contended that farmers did not really want service and, in fact, did not really need the telephone.

Mixed into this complex set of economic concerns was some measure of cultural prejudice, perceptions of farmers as short-sighted and primitive. One reads, for example, in the pages of early industry publications implicit and explicit characterizations of farmers not unlike those of then-current ethnic stereotypes. (See quotations above, for example.) Further reading between the lines of public and private documents suggests that it might have taken a great deal to convince leaders of the telephone industry that the farm population – uneducated, rustic, poor, only distantly tied to the kinds of commercial activities the industry focussed its marketing upon, and prone to use telephones in 'inappropriate' ways, such as for gabfests – formed a serious and possibly profitable market.

In sum, telephone vendors – especially the Bell companies – viewed the rural market as a marginal economic investment and a troublesome one. Whether they were correct on either count or not, and to whatever extent prejudice may have been involved, this view led the industry to neglect farmer demand except when under competitive or regulatory pressure. As cause, consequence, or both, telephone vendors may not have fully appreciated – their own advertising copy notwithstanding – the 'real needs' their instrument served in rural America, particularly the social needs. While striving to educate urbanites to need the telephone, they shunned a spontaneous market.

Thus, the farmers' massive adoption of the telephone – recall, 39 per cent of farm households versus roughly 33 per cent of those off the farms had telephones in 1920 – is all the more remarkable given the barriers they faced in addition to their late start and their peripheral location: The major vendors were largely uninterested; farmers had to overcome what most 'country life' experts considered their inherent, 'excessive individualism' in order to build cooperative systems; they needed cash, usually hard to find; and, despite upsurges of prosperity, their children and their neighbors' children were leaving the farms for the towns. Yet, the telephone was virtually a grassroots 'social movement' in most parts of rural America, for most non-Southern, economically secure farmers.

Why was the telephone so popular among American farms?

The sources of demand

First, does the fact that farm households more often had telephones than other American families actually reflect a greater popularity of the telephone on the part of farmers than townsfolk? For farmers, telephones, although classed as residential instruments, also served the farm business; perhaps they were 'only' commercial devices. Also, while fewer city people may have had telephones at home, they could use pay telephones or neighbors' telephones more easily than could rural people. Nevertheless, the evidence suggests that farm families especially had learned the 'telephone habit,' and not just, or perhaps even primarily, for business reasons.

Rural telephones seemed to be used more intensively than urban ones; and low-income farmers were willing to pay proportionally more for family use of the telephone than were town residents.[26]

Paid advertisements and publicists' stories throughout the period repeated a familiar litany of reasons why farmers should subscribe to telephone service:

> the telephone saved rural lives by being used to summon help in medical, criminal, and fire emergencies;
>
> it saved time by being used to order parts, provisions, and temporary labor from town;
>
> and it saved money by being used to call ahead for produce prices, to comparison shop, and to obtain weather forecasts; and
>
> the telephone 'banished loneliness,' provided social life, and helped keep sons and daughters 'down on the farm.'[27]

More neutral observers agreed with many of these claims. The 1907 *Census of Telephones* discusses 'the incalculable value of the telephone in the everyday life of the farm.' Some of the 'dozens' of illustrations that could be offered, its authors wrote, include: calling for current Chicago prices in order to avoid exploitation by the local grain dealer; calling into town to determine the demand for fresh produce; calling for a veterinarian or a mechanic; and stimulating social life – creating a 'sense of community.'[28]

Roosevelt's Country Life Commission reported in 1909 that poor communication was a critical problem in rural life, alongside farmers' exploitation by large corporations and their inability to cooperate with one another. The telephone was one instrument – 'already contributing much to country life and . . . capable of contributing much more' – that promised to help homesteaders learn to work together. In particular, it helped farm women's associations.[29] Ten years later, analysts of a large federal survey of farm households suggested that the telephone, together with the automobile, had indeed stimulated progress in these respects, in large measure freeing the farm family from isolation.[30]

Descriptions of the telephone's role in rural life are noticeably more specific and concrete than those of its role in urban life. The reader is left with the impression that the telephone had, in fact, more material and more important uses in farm life than in urban life, excepting perhaps for its role in urban commerce. That is, the telephone was critical in emergencies for isolated homesteads; it served the farm business (two late surveys suggest that half or more of farm household calls were for business); and it may have had more critical social uses in the countryside than in the city.

The contrast between markets is most striking with regard to these *social* uses of the telephone. The 1907 Census of Telephones argued that in areas of isolated farm houses 'a sense of community life is impossible without this ready means of communication. This is an immense boon in the life of women on the farm, who for days at a time during the planting and harvesting seasons may be left alone in the house during working hours . . . The sense of loneliness and insecurity felt by farmers' wives under former conditions disappears, and an approach is made toward the solidarity of a small country town.' The 1909 Country Life Commission and the 1919 farm survey both stressed the social role of the telephone in rural life. Testimony from rural telephonemen also dwelt on the significance, as well as the frustration, of the instrument's social role.[31] One independent company official stated in 1905:

> When we started the farmers thought they could get along without telephones . . . Now you couldn't take them out. The women wouldn't let you even if the men would. Socially, they have been a godsend. The women of the country keep in touch with each other, and with their social duties, which are largely in the nature of church work.[32]

And an academic claimed in 1912:

> No one feature of modern life has done more to compensate for the isolation of the farm than the telephone. The rural mutual lines have brought the farmers' wives in touch with each other as never before. While it is true that the visiting thus made possible sometimes interferes with the usefulness of the lines for business purposes, the telephone proves a boon to the farm household in trivial matters as well as those of greater moment.[33]

Women, it seems, were the major users of farm telephones, even for making business calls. Most farm women surveyed in 1914 argued that their key need was to break out of their isolation, and they credited the telephone, automobile, and rural free delivery with having helped them do so.

Sociability by telephone meant not only conversations with friends and neighbors, it also meant the recreational use of the telephone to listen in on others' conversations. 'James West,' pseudonymous author of a classic farming town study, stated that telephones (and automobiles) were used more as 'social conveniences' in Plainsville than as practical 'implements.'[34]

These selected testimonials receive some modest quantitative endorsement from a mass survey conducted for the Country Life Commission in 1908, asking farmers (and others) about satisfaction with various aspects of rural life. Sixty-nine percent of the 55,000 farmers were *dis*satisfied with 'social intercourse.' But 73 percent were satisfied – the highest level reported – with their postal and telephone service (asked in combination).[35]

While one must treat with some skepticism the claims of salesmen and progressive-era reformers about telephony's powers of social healing, the consensus of opinion is striking and even more is the contrast posed by the rarity of such claims for urban America.

Conclusion

The telephone industry largely ignored or underestimated the rural market. Aside from brief periods, marketing to the American farmer was minimal. Yet farmers loudly demanded the technology, had many obvious practical uses for it, and less obvious but perhaps deeper uses for it as a social tool. Surprisingly many of them, outside the South, worked hard to obtain service. This story contrasts with that of urban telephony, where the vendors worked hard at 'inventing' and selling uses for the telephone, with seemingly far less success in proportion to their effort.

The telephone, thus, may have had two forms of diffusion. One was a response to spontaneously-expressed needs in rural America, where distances between homes and settlements posed great difficulties, not the least of which was a sparse social life. People perceived their own needs for, and shaped their own uses of, the telephone. The other form of diffusion followed what some might call 'created needs' in the urban sector, where the telephone provided only marginal advantages over its alternatives – the evolving telegraph system, messenger boys, or walking. The same 1907 Census that listed many advantages of the rural telephone restricted its comments on urban telephones to specialized business activities, including grocery trade, hotel telephones, and businessmen calling the office from their vacation retreats. The contrast in social uses is particularly sharp. Town-dwellers could find companionship just outside the door, as well as convenient transportation to more distant associates. But observers generally claimed that the American farm-dweller's isolation, especially that of housewives, made the telephone a vital social instrument.[36]

Farmers were, in fact, an 'ornery' lot. Many of them demanded this technology when its vendors said it was inappropriate for them (as happened, too, in the history of rural electrification). They demanded it vociferously enough to build it themselves, to adopt it at a higher rate than did the 'natural' market of city-dwellers, and later to pester the major telephone companies through the regulatory commissions and politicians. And they insisted on using it their own way, to meet their own needs, be it for gossip or banjo concerts, rather than fit 'appropriate' uses. This story of diffusion shows more human agency, more creation of supply by consumers, more popular modification of a technology's use than would be suggested by conventional discussions of technological modernization.

Comparisons with the automobile

The argument so far is that, as a response to 'objective' needs, farmers created a demand for telephones not well understood by the telephone companies. The plausibility of this contention is strengthened by briefly examining the case of rural automobility from about 1905 to 1935.[37]

At first, rural Americans were hostile to the automobile, at least as brought to them by disruptive, littering city tourists. The *Farm Journal* argued in 1904 that 'the auto is a fad . . . This is certain.' And a rural newspaper in 1906 rejected automobile advertisements along with liquor advertisements. There were even reports of car-burnings by irate farmers. And, by and large, automobile manufacturers seemed to ignore the rural market, with Rambler placing the first advertisement in the farm press only in 1907.

But it was just about then that farmers eagerly joined the passing caravan. (Indeed, one source places the initiation of farmer enthusiasm at precisely Winter, 1907–08.) Ford's Model T and high farm prices produced a boom in rural automobility. By 1911, there were 85,000 automobiles on American farms, 25 times that number in 1920, and almost 60 times that number in 1930. According to one claim, 72 percent of cars sold in 1915 were sold to farmers. According to another, sales of 1926 models in towns of fewer than 10,000 people were almost proportional to their share of the national population – 55 percent of sales, 58 percent of Americans – which, given the poverty of rural America, indicates that small-town and rural people extended themselves to purchase automobiles. (These figures would be all the more dramatic if the southern states were set aside.)[38]

The automobile, like the telephone, became a disproportionately rural technology. And, like the telephone, it was especially common across the vast stretches of the Plains States. Their residents outpaced, quickly and by a wide margin, those of the urban East in automobile diffusion. By roughly 1910, there were more motor vehicles per capita in North Dakota, South Dakota, Nebraska, Kansas, and Idaho than in Connecticut, Rhode Island, New York, New Jersey, and Pennsylvania (about 15 per thousand residents versus 10; in 1927, the difference was about 250 to 170). These differences emerged in spite of more affluence and more commercial demand for motor vehicles in the urban states. The proportion of farms reporting an automobile was 31 and 58 in 1920, and 1930, respectively.[39]

Reports of rural Americans' enthusiasm toward the automobile during the 1910s and 1920s were frequent. Neighbors often gathered for 'Good Roads Days,' collective road-building amidst festivity and feeding. 'Ford Days' were set aside at county fairs for stunts using the Tin Lizzies. Farm women especially welcomed the automobile as an antidote to rural monotony. One explained to an interviewer that her family bought a Model T before purchasing indoor plumbing: 'Why you can't go to town in a bathtub!'[40] More detailed evidence underlines the enthusiasm. In a 1922 survey, white farm families reported spending an average of $110 a year on non-business use of automobiles alone. A USDA survey in 1935–36 showed that farmers spent more money on automobile travel than did low-income town residents.[41]

Why did American farmers adopt the automobile so eagerly? As with the telephone, there were many obvious practical advantages. It could be used for the business of the farm, such as hauling goods to market, bringing day laborers in from town, and quickly obtaining spare parts. Farmers were more likely than townspeople to buy automobiles for business reasons. And it could be used for chores at home (set up on a jack, even to turn a washing machine). An automobile – even an early Ford – was more reliable, efficient, faster, safer, healthier, and probably cheaper than a horse. Automobile travel was considerably more flexible than the railroad for long trips and also opened up new opportunities, such as touring, to formerly homebound farmers.[42]

The automobile, like the telephone, was used socially as well as practically. Only about a fifth of farmers' automobile expenses in the late 1930s could be charged to business use. The rest was for 'family living.' As noted earlier, 'James West' found that Plainsville residents used their cars more for 'social convenience.' There were claims, as with the telephone, that it kept rural women sane by breaking their isolation. That the automobile increased social contacts was argued not only by automobile publicists, but also by other observers, such as the 1909 Country Life Commission, in its call for better roads, and the 1919 farm household survey discussed above. A 1915 mail survey of farmers came back to the USDA showing that 'women and men in almost every state held that the greatest service the department could render to the farm population would be systematic improvement of country roads [for automobiles].' Some contemporary researchers thought that the multiplication of extended contacts came at the cost of reduced ties with immediate neighbors, but there seemed to be a consensus that total social involvement increased, as evidenced, for example, in more and larger regional social organizations.[43]

We can see a familiar contrast to the urban market. Many, in and out of the industry, saw the automobile as essentially a recreational item – as late as the 1930s by the authors of *Middletown*. Few city residents bought it for business reasons or spent much money on automobile travel for such purposes. And only a minority commuted to work by car. A Pittsburgh survey of 1934 showed that half the metropolitan households owned automobiles, but in only 20 percent did someone use it to commute to work. It may be that the automobile required more 'creating of needs' and more intense advertising in urban places (what one contemporary sociologist called creating 'unrest' and 'a channel for its release'), while rural diffusion arose more out of obvious, concrete needs – as perhaps did the rural telephone.[44]

And, as in the case of the telephone, the vendors of automobiles were caught off guard by rural demand. One report claimed that a shortage of 15,000 automobiles appeared in 1908 because of unanticipated farmer demand. Manufacturers appeared to move gingerly into the market, with the exception of Henry Ford. Although he shunned advertising *per se*, Ford aggressively marketed the vehicle to farmers. By 1913, every town of 2,000 residents or more had a Ford dealership. By 1930, 65 percent of his dealers were in rural areas, as opposed to 46 percent of his competitors'. The other manufacturers could not match Ford's rural appeal, but they did follow him into farm territories. Although perhaps initially unexpected, the rural market was eventually served and continued to grow throughout the entire period.

Conclusion

Although the contrast between urban and rural selling of the automobile is not as great as that with the telephone, there are parallels. The industry generated heavy, favorable, and often subsidized press coverage to sell to the urban affluent. The rural public was relatively neglected, at least at first. One rural spokesman claimed that 'before the automobile went spinning through his farm, [the farmer] didn't realize how far from his neighbors, how isolated was his life.'[45] This theory does not seem to fit history; farmers responded to the automobile, perhaps because the needs it satisfied were concrete, meaningful, and long-standing – not the least among these needs being the facilitation of social life. Yet, rural Americans demanded the automobile with an intensity well exceeding that of the more affluent city-dwellers. Unlike the telephone, there was a quicker, fuller, more lasting response by the industry.

Discussion

Farm telephony stagnated and then actually declined in the 1920s and 1930s – in absolute numbers as well as in proportion of farms with a telephone. Roughly one million fewer telephones were connected in American farm homes in 1940 than in 1920, down from 39 percent to 25 percent. I discuss why telephone service shrank in rural America elsewhere.[46]

Eventually, most American farmers did acquire telephones and automobiles; technological modernization may be said to have swept them up. But the story of rural telephony – and to some extent, of rural automobility, both central in the constellation of 'modern' implements – challenges simplistic images of an inevitable tide of modernization sweeping from the urban centers over the backward countryside.

First, we see that the course of telephones' spread was contingent on businessmen's perceptions, decisions, and prejudices, on legal contexts, such as patents and regulations, and on political circumstances in different eras. The contextual issue is highlighted by contrasting the American experience to that of Western Europe. There, state authorities typically served rural areas quickly with public telephones, but were late in providing farmers with private telephone exchanges. (In Scandinavia, however, rural cooperatives sprung up plentifully, as they did in the American Midwest.)[47]

Second, we see that, at least for a period, the rural hinterland outpaced the urban core in embracing a technological innovation – and this despite numerous disadvantages, among them the general conviction of telephone suppliers that farmers were too backward to demand telephones.

And third, the best understanding of why this reversal occurred suggests that farmers, individually and collectively, saw reasonably well and reasonably early the material and social advantages of the telephone, demanded it, and in most regions supplied it themselves, if necessary. City-dwellers, although more affluent and more sought-after by vendors, were relatively more resistant. Moreover, farm families used this tool in ways not recommended by its sellers, most notably for sociability and entertainment. Human – here, consumer – agency was paramount in modernization.

This case suggests, more generally, that we need to ask not, How does technology 'impact' people? or, How does modernisation alter social life?, but instead ask, How do people come to choose, use, and adapt a technology?, and, How do people manipulate new technologies in constructing what we after the fact call 'modern' ways of life?

Notes

1 See C.S. Fischer, 'Studying Technology and Social Life,' in M. Castells (ed.), *High Technology, Space, and Society* (Beverly Hills, 1985): 284–301.
2 I. de S. Pool, 'Introduction' to Pool (ed.), *The Social Impact of the Telephone* (Cambridge MA, 1977): 4. These issues are pursued in two studies of the telephone: C.S. Fischer, '"Touch Someone:" The Telephone Industry Discovers Sociability,'*Technology and Culture*, In Press; and idem, 'Technology's Retreat: The Decline of Rural Telephony, 1920–1940,' Paper presented to the American Sociological Association, 1985.
3 On urban-to-rural diffusion, see, for example, P.O. Pedersen, 'Innovation Diffusion Between and Within National Urban Systems,' *Geographical Analysis*

3:201–54; C.S. Fischer, 'Toward a Subcultural theory of Urbanism,' *American Journal of Sociology* 80 (1975): 1319–41; and on diffusion generally, E. Rogers, *The Diffusion of Innovation*, 3rd Edition (New York, 1983).

4 For general histories of the telephone, see G.W. Brook, *The Telecommunications Industry* (Cambridge, MA, 1981); J. Brooks, *Telephone: The First Hundred Years* (New York, 1976); N.R. Danielian, *AT&T: The Story of Industrial Conquest* (New York, 1939); M.M Dilts, *The Telephone in a Changing World* (New York, 1941); R. Garnet, *The Telephone Enterprise* (Baltimore, 1985); J.E. Kingsbury, *The Telephone and Telephone Exchanges* (London, 1915); H.B. MacMeal, *The Story of Independent Telephony* (Chicago, 1934); and I. de s. Pool, ed., *The Social Impact*, op. cit.

5 Statistics from U.S. Bureau of the Census (USBOC), *Historical Statistics of the United States*, Bicentennial Edition, Part 2 (Washington, 1975), pp. 783–84.

6 This typology was used by the first census of telephones, conducted in 1902, and published in USBOC, *Special Reports; Telephones and Telegraphs 1902* (Washington, 1906): 7.

7 Ibid.; Federal Communications Commission, *Proposed Report; Telephone Investigation* (Washington, 1938): 147; idem., *Investigation of the Telephone Industry in United States* (Washington, 1939): 132–33. AT&T has always challenged this interpretation; see, e.g., American Telephone and Telegraph, *Annual Report of AT&T 1909* (New York, 1909): 26–28.

8 See, for example, H.N. Faris, 'Some Lessons Learned From the Past,' *Telephony* (22 May, 1926): 17; and T.N. Gary, 'The Independents and the Industry,' *Telephony* 90 (13 March, 1928): 14–18.

9 Relatively little is known about the history of rural telephony. For example, two pages on rural telephony are indexed in Brooks, op. cit., and three in Pool (ed.), *Social Impact*. This narrative is pieced together from fragmentary sources. Among general references are the various *Censuses of Telephones*; the pages of *Telephony*; and corporate and association histories in the files of: AT&T Historical Archives (New York), the Museum of Independent Telephony (Abilene, Kansas), the San Francisco Telephone Pioneer Museum, Illinois Bell Communications (Chicago), and Bell Canada Historical (Montreal).

10 The 1902 and 1907 statistics are from USBOC, *Special Reports: Telephones: 1907* (Washington, 1910): 24, 16. On the likelihood of a 1907 undercount, see 'Report of Secretary Ware,' *Telephony* 18 (11 Dec., 1909): 651–54. After that census, the Bureau essentially gave up trying to distinguish rural and urban telephones and, in lieu, used size of company. The 1912 estimates are from the *Wall Street Journal*, quoted in *Pacific Telephone Magazine 6* (Oct., 1912): 12.

11 The 1902 estimate assumes that all rural telephones counted in the USBOC, *Telephones and Telegraphs, 1907*, were on farms. See, also, S. Lebergott, *The American Economy* (Princeton, 1976): 351–55. The national estimate is based on various indirect sources. The 1912 farm estimate is from U.S. Department of Agriculture, *Income Parity for Agriculture*, Part III, Section 3, Preliminary (Dec. 1938): 5. The 1912 national estimate is calculated from statistical retrojections of later figures. The 1920 farm estimate comes from USBOC, *Census of Agriculture, 1950*, Volume II, General Report (Washington, 1952): 211, while the national figure can be found in idem., *Historical Statistics*: tables R1–R12. In 1920, the states where farmers were most likely to have telephones were those with relatively more farm owners than farm tenants, relatively few black farmers, valuable land, high density (i.e., the most sparsely settled states tended to be low in farm telephony), and outside the deep South. See C.S. Fischer, 'Technology's Retreat' op. cit. On Canada: R. Pike and V. Mosco, 'Canadian Consumers and Telephone Pricing,' *Telecommunication Policy* (March, 1986); 17–320.

12 There is, for example, the case of Marysville, California, in the 1890s. The

local constable had taken a correspondence course in telephone engineering and then wired up neighboring farmers all the way to the town line. Although the Bell company had a franchise in Marysville that shut out the upstart, it was compelled by this threat to its town market to provide the farmers with Bell service at 25 cents a month (PT&T News Bureau Files, Pioneer Museum).

13 J.B. Mulrooney (Commercial Engineer, New England T&T), 'Some Aspects of Rural Telephone Service,' *Telephony* 112 (9 Jan., 1937): 14–16.

14 In 1893, New England T&T charged $3.00 per quarter mile after the first mile from the central exchange – Mulrooney, op. cit.

15 Report of G.B. Bush (General Commercial Superintendent, PT&T) to E.C. Bradley (Vice–President, PT&T), 21 Nov., 1908, passed on to F.A. Pinckernell (AT&T), 25 Nov., 1908, in 'Rural Telephone Service – 1907–1910,' Box 1363, AT&T Archives.

16 'Minutes of the Advertising Conference, Bell Telephone System . . . 1916,' in Box 1310, AT&T Archives. The major exception to this passivity seemed to be Southwestern Bell.

17 A 1914 advertisement showed a drawing of a farmer. Its text read: '"The city is as near as my telephone . . ." That's the thought we had in mind in the development of Rural Bell Telephone Service' (Advertising folder, Pioneer Museum). A 1915 advertisement, 'Neighborizing the Farmer,' states that '[o]ne of the most significant facts of our telephone progress is that one-fourth of the 9,000,000 telephones in the Bell System are rural . . . Today, the American farmer enjoys the same facilities for instant, direct communication as the city dweller . . . The Bell System has always recognized rural telephone development as an essential factor of Universal Service' (Ayer Collection, National Museum of American History, Smithsonian Institute). [. . .]

18 G.F. Wonbacher, 'Proper Development of the Rural Telephone,' *Western Telephone Journal* 12 (July, 1908): 242.

19 On 'frightened', see B.G. Hubbell (President, Frontier Telephone Company, Buffalo), testimony to State of New York, *Report* of the Committee of the Senate and Assembly Appointed to Investigate Telephone and Telegraph Companies (Albany, 1910): p. 791. On farmers' ignorance, see 'Growth of Farmer Lines in Iowa,' *American Telephone Journal* (13 Sept., 1902): 150–51; and J.W. Andrews, 'The Telephone in Farm Life,' ibid., (6 Dec., 1902): 343. On entertainment: 'In the rural districts telephones were installed in a great many instances only to satisfy the young people of the family. The first years I was around collecting up the rents . . ., I would often hear, from the old man who footed the bills, the remark: "That's a costly plaything for the young ones"' (T. Holein, 'Rural Telephone Service,' *Telephony* 17 [6 Feb., 1909]: 158). On 'Don't be afraid:' W.H. Barker, 'Operating Rural Lines,' *Telephony* 19 (5 March, 1910): 286.

20 A.B. Fishback, 'Our Rural Friends,' *Western Telephone Journal* 9 (March, 1907): 88.

21 L.P. Boies, 'Concerning Farmer Lines,' *Western Telephone Journal* 12 (Feb., 1908): 6.

22 The following item from the *Algona (Iowa) News* probably describes an extreme case: 'Gossip over a party telephone line, alleged personal remarks about neighbors, and too persistent listening by people on the line have resulted in a battle, caused a serious feud in Kossuth County and doubtless will end in a long drawn-out battle in the courts.' The story explains that people overheard gossip about themselves on the party line. 'The trouble became so pronounced that it was decided to hold a meeting of the subscribers . . . But this meeting, instead of helping matters, had the reverse effect. Following heated arguments and acrimonious retorts the gathering ended in a free-for-all fight . . . Lifelong friendships have been broken, relatives have become strange,

and it is said that a dozen law suits will result from the feud' (reprinted in *Telephony* 15 [March, 1908]: 230).

23 Reprinted, with endorsement, in *Western Telephone Journal* 12 (July, 1908): 258.

24 'Advertising Conference – Bell System – 1916,' Box 1310, AT&T Archives: 36

25 Rural areas were largely ignored, the 1907 *Census* argued, because the 'extraordinary demand' for the telephone in the cities fully occupied the energies and capital of the Bell companies (USBOC, *Telephones: 1907:* 76). Given the clear slowdown in telephone expansion during the 1880s – attributed by some to technical problems that were predominantly urban, the electric railways – and in other periods, time and energy could have been released to serve rural customers, so that may not be a sufficient explanation.

26 From 1902 through 1912, an average of 8.4 calls per telephone were made in the west north-central states, versus 5.1 in the middle Atlantic states; from 1927 to 1932, the rates were 6.1 versus 4.6 – BOC, *Telephones and Telegraphs . . . 1912* (Washington, 1915): 32, and idem, *Census of Electrical Industries 1932: Telephones and Telegraphs:* 21.

27 These claims appear everywhere in the small rural telephony literature, such as the Bell System's 'Neighborizing the Farmer' advertisement; a circular of the North Electric Company of Ohio, reprinted in *Telephony* (9 [April, 1905]: 303]; and a booklet on how to form mutual companies by the Stromberg-Carlson Telephone Manufacturing Company, *Telephone Facts for Farmers* (Rochester, N.Y., 1903 [Warsaw Collection, National Museum of American History, Smithsonian Institution]); newspaper stories (quite likely planted) such as 'The Telephone on the Farm,' *New York Sun* (reprinted in *Telephony* 6 [November, 1903]: 383–84); and P.C. Henry, 'A Cooperative Telephone,' *Country Life in America* 24 (May, 1913); AT&T-subsidized books such as H.N. Casson, *The History of the Telephone* (Chicago, 1910) and A. Pound, *The Telephone Idea: Fifty Years After* (New York, 1926); and *Telephony* articles such as 'Rural Telephones,' (8 [July, 1904]: 34–35), F. De Land, 'Eliminating Isolation from Farm Life,' (9 [March, 1905]: 255–58), and R.F. Kemp, 'Telephones in Country Homes,' (9 [June, 1905]: 432–33).

28 USBOC, *Telephones: 1907:* 74–81.

29 U.S. Congress, Senate, *Report of the Country Life Commission*. 60th Cong., 2nd Sess., 1909. S. Doc. 705: 93.

30 F.E. Ward, *The Farm Woman's Problems*. USDA Circular No. 148. (Washington: U.S. Government Printing Office, 1920). See, also: Butterfield, op. cit., and F.W. Card, 'Cooperative Fire Insurance and Telephones,' in L.H. Bailey (ed.), *Cyclopedia of American Agriculture*, Volume IV, 4th Edition (New York, 1912): 303-06.

31 USBOC, *Telephones: 1907*: 77–78; U.S. Senate, *Country Life*, op. cit.; 1919 survey: Ward, op. cit.

32 Quoted in Kemp, op. cit.: 433. A 1909 article claims that '[t]he principal use of farm line telephones has been their social use . . . The telephones are more often and for longer times held for neighborly conversations than for any other purpose.' It goes on to stress that subscribers valued conversation with anyone on the line (G.R. Johnston, 'Some Aspects of Rural Telephony,' *Telephony* 17 [8 May, 1909]: 542).

33 Card, op. cit.: 305. Of course, there was some dissent. Chicago sociologist R.D. McKenzie endorsed the view, in 1928, that the telephone had helped alienate country households (in A.H. Hawley [ed.] *Roderick D. McKenzie on Human Ecology* [Chicago, 1968]: 168).

34 J. West (C. Withers), *Plainsville, U.S.A.* (New York, 1945): 10.

35 O.F. Larson and T.B. Jones, 'The Unpublished Data from Roosevelt's Commission on Country Life,' *Agricultural History* 50 (Oct., 1976): 583-99.

36 On advertised uses, see C.S. Fischer, '"Touch Someone,"' op. cit.

37 General sources on the history of the automobile in rural America include: M.L. Berger, *The Devil Wagon in God's Country: The Automobile and Social Change in Rural America, 1893-1929* (Hamdon, CT, 1979): J. Interrante, 'You Can't Go to Town in a Bathtub; Automobile Movement and the Reorganization of Rural American Space, 1910-1930,' *Radical History Review* 21 (Fall, 1979): 151-68; N.T. Moline, *Mobility and the Small Town, 1900-1930: Transportation Change in Oregon, Illinois* (Chicago: University of Chicago Department of Geography, Research Paper No. 132, 1971); and R.M. Wik, *Henry Ford and Grass-Roots America* (Ann Arbor, 1972).

38 See, besides general sources cited above: L.J. Carr, 'How the Devil-Wagon Came to Dexter,' *Social Forces* 11 (October, 1932): 64–70. The precise timing is in A.P. Haire, 'The Farmer and His "Benzine Buggy,"' *Printers' Ink* 62 (25 March, 1908): 10-15. C.M. Harger explains that the vast distances, decent roads, and flatness of the plains states especially invited automobiles ('The Automobiles in Western Country Towns,' *World Today* 13 [Dec. 1907]: 1277-79).

39 State figures were calculated from motor vehicle registration data in Public Roads Administration, *Highway Statistics Summary to 1945* (Washington, 1947). The figures for the years before circa 1915 are often based on estimates. The farm data are from USBOC, *1940 Census of Agriculture, General Report, Statistics by Subjects* (Washington): 512.

40 Quotation from Interrante, op. cit. See also Wik and Berger, op. cit.

41 1922: E.L. Kirkpatrick, *The Farmers' Standard of Living*, Bulletin No. 1466, U.S. Department of Agriculture (Washington, 1926): 24. 1935: D. Monroe et al, *Family Expenditures for Automobile and Other Transportation, Five Regions*. Consumer Purchases Study: Urban, Village, and Farm. Bureau of Home Economics, Misc. Pub. 415. (Washington, 2941): 2, 39ff.

42 Forty-two percent of farmers versus about 20% of village and small town respondents in the 1935-36 survey reported above cited business use (Monroe et al., op. cit.: 10-11, 55). On chores, see Wik, op. cit.: 29-32.; and on travel: Moline, op. cit.; W.J. Belascoe, *Americans on the Road: From Autocamp to Motel, 1910-1945* (Cambridge, MA, 1979).

43 Expenses: D. Monroe et al, op. cit.: 34–36. Plainsville: West, op. cit.: 10. On women: see Wik, op. cit.: 25. Calls for roads: *Country Life*, op. cit.; Ward, op. cit.; USDA, *Social and Labor Needs of Farm Women,* Report 103 (Washington, 1915): 66. A Vermont village history also points to the link between automobility and sociability – S.B. Clough and L. Quimby, 'Peacham, Vermont,' *Vermont History* 51 (1983): 5-28. On neighbors versus wider ties, see J.H. Mueller, 'The Automobile: A Sociological Study,' Ph.D. Diss., Department of Anthropology and Sociology, University of Chicago, 1928; and F.R. Allen, 'The Automobile,' in F.R. Allen et al., *Technology and Social Change* (New York, 1957): 107-32.

44 R.S. Lynd and H.M. Lynd, *Middletown* (New York, 1929). Pittsburgh: J. Tarr, *Transportation Innovation and Changing Spatial Patterns in Pittsburgh, 1850-1934*, Public Works Historical Society Essay No. 6 (Chicago, 1978). And on 'unrest': Mueller, -op. cit.: 33.

45 Quoted by Wik, op. cit.: 208.

46 C.S. Fischer, 'Technology's Retreat: The Decline of Rural Telephony,' op. cit.

47 On crossnational comparisons, see A.N. Holcomber, *Public Ownership of Telephones on the Continent of Europe* (Cambridge, MA, 1911); also discussion in C.S. Fischer, 'Technology's Retreat.'

4
Science, Technology and Welfare

4.0 Introduction

In the first half of the volume, we failed to identify any specific type of technology as the prime mover of social change, encountering instead objections which might be levelled at any other such choice. In the latter half of the volume, in order to advance the discussion, we consider the effects of ensembles of technological innovations, and of a technocratic ideology.

Although our theme of causality recurs throughout this chapter, its organizing principle of human welfare brings to the fore related questions about technology and values. In a period when an ideology of scientific and technological progress prevailed, did such innovation in fact improve the quality of life? And if there were major improvements, could these always be attributed to technological advances? As we might expect, contradictory answers emerge from the readings below. F.B. Smith's particular conclusion is that before 1941 and the introduction of penicillin, scientific medicine had little impact on maternal mortality in Britain; indeed, a pregnant woman's chances were better elsewhere than in hospital. Anthony S. Wohl presents a generally grim picture of industrial disease in Victorian Britain, but finds no simple correlation between advanced technology and health hazards; the fact that large factories could be more easily regulated, for example, meant that their inmates generally fared better than employees in small workshops. Eric Ashby and Mary Anderson make the point that from the 1880s, the infamous London smog was due less to heavy industry than to the traditional domestic fire. Moreover, the technological means to abate the smoke were available in the form of closed stoves, but these met with 'sociological' barriers.

This check by consumer preference on a technological solution to a social problem may be set against the positive part of consumer choice in the diffusion of telephony, described in the previous chapter by Claude S. Fischer. Further comparisons with Fischer arise from E.J.T. Collins' article on factory-produced bread and cereals: again we find the fulfilment of corporate objectives checked by consumer preference, this time conservatism in diet. A more general lesson can be drawn from Collins: in the case of ready-to-eat cereals, technological innovation did not itself force social change, but rather filled a niche created by much broader movements in social patterns. In the extract by Nicholas Bullock and James Read, a vital area of human welfare is seen to be relatively untouched by technology in the popular 'hardware' sense. Improvements in housing conditions in Europe, at least up to 1914, have nothing to do with the new construction methods described in the first two readings of Chapter 1, though much to do with the efforts of sanitary reformers.

The sanitarians were, however, clearly motivated by a belief that social problems could be eliminated by 'scientific' methods, in the form of planning, regulation and inspection. They were engaged in a kind of social engineering, and ironically shared with their eugenist opponents an ultimately technocratic ideology. The eugenists, for their part, invoked the science of heredity in support of their belief that society could only be improved if its 'unfit' members, instead of being encouraged by public health measures, were actively prevented from reproducing themselves. Donald MacKenzie's article argues that in the British context, the eugenists' appeal to scientific fact expresses the interests of the new professional classes.

4.1 E.J.T. Collins, *The 'Consumer Revolution' and the Growth of Factory Foods: Changing Patterns of Bread and Cereal-Eating in Britain in the Twentieth Century*, 1976

Since 1900, the demand for cereal foods, which historically were a mainstay of human dietary, has been affected by changes in real incomes and by the forces of the 'consumer revolution' (see Table I).

Table I:
The food share of consumer expenditure 1900–65[1]
(percentage at current prices)

1900–09	*1910–19*	*1920–29*	*1930–39*	*1940–49*	*1950–59*	*1960–65*
28.5	30.9	31.4	27.4	23.5	26.7	23.7

Cereals moving into consumption in the UK 1900–70
(in lbs. per head per annum)

1909–13	*1924–28*	*1934–38*	*1941*	*1944*	*1950*	*1960*	*1970*
237	214	210	257	253	223	181	160

Cereal expenditure as a percentage of total food expenditure

1900–09	*1910–19*	*1920–29*	*1930–39*	*1940–49*	*1950–59*	*1960–69*
16.0	17.0	14.6	14.6	17.9	16.4	15.3

The pattern in 1900 – wheaten flour and white bread, with cakes and biscuits semi-luxuries, the output of manufactured cereal foods small and unimportant and oatmeal porridge popular only in the higher income groups and in Scotland – was tedious and unexciting compared with a

Source: Derek Miller and Derek Oddy (eds), *The Making of the Modern British Diet*, London: Croom Helm, 1979, pp. 26–43.

century earlier when cereals were eaten in many different forms – wet, dry, hot, cold, baked, griddled – and when there was much local and regional variety in the colour and composition of the household loaf.[2] Since 1900 the trend has been one of increasing diversification, characterised by a switch from the cheaper more traditional foods to the costlier and more highly formulated foods. The overall result has been a fall in the proportions of both the weight and the value of bread and flour to total cereal consumption.

Bread

The price and income elasticity of demand for bread has been lower than for most other major categories of foodstuffs, and since 1900 it has become an 'inferior good', in the sense that consumption has fallen as real incomes have risen. The reasons are well known and require no repetition beyond emphasising the importance of changes in eating habits (e.g. the decline of the 'packed lunch' and 'afternoon tea'), of the demand for lighter and more savoury foods, and the many social factors, grounded in the 'consumer revolution', which streamlined the domestic routine.[3] Bread consumption between the wars may have levelled off because of the increasing popularity of the sandwich, and although bread-eating reached new peaks during and immediately after the Second World War, consumption had already fallen below pre-war levels by 1953, and thereafter sank rapidly. But although *per capita* consumption has fallen, the industrial production of bread is still very much higher today than in 1920, as a result partly of population growth and partly of the decline of home-baking (in 1920 loose flour represented about 25 per cent of wheaten flour sales compared with 14 per cent in the late sixties).[4] The greater expansion, however, was in flour confectionery, whose production rose from 35 per cent of bakery sales in 1924 to 46 per cent in 1954.[5]

These trends must be seen against the broader background of structural change in the bakery industries. A major development which began about 1930 was the displacement of the small master baker by the large plant bakery. In 1937 plant bakeries were responsible for only 10–12 per cent of national bread output, but by 1953 the proportion had increased to 35–40 per cent.[6] The economics of plant bakeries were those of mass production, economies of scale and vertical integration into the retail trades.[7] Multiple-shop bakeries began in the nineteenth century with the co-operative movement and by 1939 over 30 per cent of bread sales were channelled through about 2,400 shops owned by just 79 firms.[8] Allied Bakeries, founded in 1935 by Garfield Weston, was the first private firm to distribute bread nationally, and by 1938 it owned directly or through its subsidiaries 28 bakeries and 217 shops.[9] [. . .]

The chief limiting factor in the growth of the bread market has been falling *per capita* consumption. Advertising has had little effect on demand other perhaps than to redistribute brand shares, either in the 1930s or, more recently [. . .]

Perhaps the most severe constraint, even within the fairly narrow limits of product design and manufacture, has been the strength of consumer preference for a single product, the white wheaten loaf, which comprised in 1900 over 95 per cent and in 1970 still over 80 per cent of bread consumption by weight.[10] Speciality and patent breads first appeared in the 1880s and by 1912 a few, such as Hovis (reinforced wheat germ), Daren (wheat, ryemeal and wheat germ), Kermode's (wheat and maize), Triagon (wheat, maize and rice) and Malt Loaf, had acquired a regional or national reputation.[11] The range now includes 'high-protein' and 'low-calorie' breads and also crispbread (strictly a biscuit and a bread substitute) which was introduced between the wars as a slimming food (the firm of Ryvita opened its Birmingham factory in the 1930s). The main competition, however, was between white breads and brown breads, which latter, made from low-extraction flours, were virtually extinct in Britain in 1850, but were reintroduced, primarily as health foods, in the last quarter of the nineteenth century. Yet after many years of exhortation and advice by food scientists and health reformists as to its superior nutritive value, the brown bread share of the national bread market is today still less than 10 per cent.[12]

The difficulties in breaking down a consumer preference of this order can be illustrated, by way of digression, from the history of the firm of Hovis Limited which, since the 1890s, has been the UK market leader in brown bread.[13] Hovis is a firm of millers manufacturing a patent flour which is sold to the bakery trade. The special feature of Hovis flour is its super-high concentration of wheat germ which in ordinary flours is removed in the milling process. [. . .]

The product was locally successful and, in 1896, it was decided to increase production and expand the market into the larger centres of population, until by 1939 Hovis controlled between 7 and 8 per cent of national flour output.[14] The problems though, were not of supply, but of demand – of winning the confidence of the public and the support of the bakery trade. Thus, it was necessary to campaign simultaneously at each main link in the chain of manufacture and distribution. The baker was won over by the offer of a nationally advertised branded product together with a wide range of supporting services. In 1895 a printing plant was set up at Macclesfield to produce leaflets, billheads, advertisement cards, paper bags, cardboard boxes, bread bins and delivery baskets, all of which carried a Hovis advertisement, for sale to the trade at competitive prices. In 1905 these activities were extended to include the design and manufacture of signs and sign boards for bakers' vans and barrows. The van departments, the first of which was established in London in Buckingham Palace Road, began later to undertake van repair work and actual van-building, which not only increased the company's share of in-trade advertising but was also a valuable source of income between the wars when door-to-door delivery by motor lorry and electric van became the vogue. The Company also provided the baker with special baking tins, each with the Hovis imprint, which he was obliged to use when making Hovis bread, and which was a successful first attempt, prior to the introduction of the wrapped loaf, at brand-naming. As well, it organized, from the 1890s, a programme of regional and national baking competitions to improve standards of Hovis bread-making and to popularize the art and the product.

In the approach to the consumer, the aim throughout has been to overcome the popular prejudice against the brown loaf by emphasizing product benefits. In the early years the goal was to acquire a sound reputation with a public that was sceptical of patent foods and wary of adulteration. Thus early advertisements made considerable play of royal patronage, of awards and diplomas for quality and purity, and of the need to beware cheap imitations. Hovis was depicted as a health food – as a cure for indigestion and source of 'good bone, brain, flesh and muscle' – conveyed by the messages 'Hovis builds strong men', 'Keep the wolf from the door' and 'England's pride and England's glory'. Initially, press advertising was centred on quality journals and magazines like *Strand*, *Black & White* and *Illustrated London News*, but in 1906, in a bid to broaden the base of the market, several advertisements, ranging to full front-page spreads, were placed in national daily newspapers, such as the *Graphic* and *Daily Mail*.

By 1914 the trade mark had become a household word, and a base had been built from which to launch the intensive promotional campaigns of the inter-war years. Hovis employed simultaneously several different techniques, directed at different sections of the market and at different susceptibilities. Brash 'cartoon-ads' – in 1926 even a cartoon film – competed with pseudo-scientific advertisements aimed at creating in the mind of the consumer a dietary problem which only Hovis could resolve. 'Do you', asked one advertisement, 'buy a loaf by its colour or by its food value?' Hovis identified itself more positively with the *sans corpa* through its association with the 'Fitter Britain Movement' and with Thomas Inch, the muscle-man, and author of many books and pamphlets on health, diet and body-building. In-store and in-trade advertising were carried on under the cloak of services to the bakery trade. Outdoor advertising was liberally spread over vans, buses, railway stations, roadside hoardings and at cafes and restaurants. The growing public interest in hiking, motoring and cycling brought in its wake a rash of country tea-shops, and with them, 'Teas with Hovis' signs, and special tea-sets, which were a Hovis novelty in 1900, but found a perfect niche in the mock-Tudor world of the 1930s. [. . .]

The firm of Hovis has been a major force in the modern food industry, but at no point has its share of the total bread market exceeded 5 per cent. Up to the Second World War brown bread was a predominantly upper-income food; a 1936 survey records that average consumption in income groups A and B was more than three times greater than in income groups C and D.[15] Since 1950 the gap has closed, to the point – which raises interesting speculations as to future trends in demand – that consumption is now highest in income group D households, and highest of all among old age pensioners.[16]

The origins of breakfast cereals

Ready-to-eat (RTE) breakfast cereals had the style and versatility that the bread foods had not. They originated in the United States in the second half of the nineteenth century.[17] They were developed to meet the needs of vegetarian groups like the Seventh Day Adventists who, at Battle Creek, Michigan, were already experimenting with cereal-based foods in the 1850s, and later, of dietary reformists who were an important market for patent foods. The concept and formulation of breakfast cereals owed much to the work of Dr John Harvey Kellogg, the Director of the 'medical boarding house' (later the Sanitarium) at Battle Creek, where he researched into the 'dietary problem' and the development of so-called 'natural foods'. In the 1860s he formulated 'Granola', the first ready-cooked breakfast cereal food, made from a mixture of wheat, oatmeal and maize baked in a slow oven until thoroughly dextrinised. In 1896 he produced 'Granose', the first flaked cereal food, made of wheat, and three years later, by the same process, the inimitable corn flake. Kellogg also helped to popularise the science of applied dietetics and to stimulate experimental work elsewhere on patent cereal foods. By the 1890s many minds were bent to the same problem, and the decade around 1895 saw in quick succession the development of most of the basic types of pre-cooked cereal and manufacturing process (i.e. flaking, toasting, puffing and extrusion). Shredded Wheat was invented in 1892 by Henry D. Perky of Denver; Grape Nuts in 1898 by Charles W. Post (an ex-patient of Kellogg's at the Sanitarium); Puffed Wheat in 1902 by Alexander Anderson, and Toasted Oat Flakes, again in the 1890s, by the Beck Cereal Company of Detroit. These discoveries plunged the American Midwest into something of a breakfast cereal boom, but of the many inventions, few survived the outbreak of the Great War. By 1920 the less efficient producers had been forced out of business, and control of the market had passed into the hands of a few large firms headed by Kellogg, the Force Food Company of Buffalo, the Shredded Wheat Company of Niagara Falls (now part of National Biscuit), Quaker Oats, Armour Grain of Chicago (now the Ralston Purina Company) and the Postum Cereals Company (now the Post Division of General Foods).

The factors underlying the successful intrusion of breakfast cereals into the North American and British dietary patterns were complex. Though conceived originally as health foods, their success in the longer run was due to other, more dynamic forces, such as rising real incomes, changing patterns of work and employment and the overall quickening tempo of twentieth-century living, which created a niche for lighter and quicker foods. There were other important influences. Special opportunities for product design and innovation were afforded by the uniquely twentieth-century phenomenon of the children's market, at which, after 1920, the main promotional effort was directed. The numbers of working wives were growing, and convenience foods, especially in the early morning, became a part of the domestic routine.

Many of the early pioneers of the breakfast cereals industry had been motivated by an interest in dietetics, but it was vigorous salesmanship that convinced the public at large that breakfast cereals were a utility food and not just a 'health food oddity'. Charles Post, the inventor of Grape Nuts and Post Toasties, was the first to apply the symptom-inducing advertising techniques used by the makers of patent medicines. In pioneering the art of 'selling health foods to well people', he widened the appeal of scientific eating and anticipated the important role that the food industry was to play in stimulating (and focusing) public interest in health and nutrition. W.K., brother of John Harvey, established the Kellogg Co. as world leader through skilful advertising and single-minded concentration on the large popular markets in the American cities.[18] The year 1906 saw the 'Wink at your Grocer' campaign; 1907, the first 'Sweetheart of the corn' advertisement; 1912 the erection in Times Square of the world's largest illuminated sign, and the 1930s, in the face of depression and falling prices, the doubling of advertising expenditures, and expansion overseas. Brand image was embodied in catch-phrases like the 'Sunshine breakfast', 'Just a little bit better', 'The road to Wellville', and, most enduringly, 'The original has this signature'. The US breakfast cereals industry pioneered the wax-tite container and dynamic packaging. It evolved new techniques of product development, test and consumer-marketing and in-store promotion. Since the 1920s the trend in the industry has been towards higher levels both of product concentration (in 1950 five companies accounted for over 95 per cent of RTE sales), and of conglomeration (incorporating other food manufacturing interests such as pet foods, groceries and poultry).[19] In short, the achievement was the wholly formulated, socially engineered product, and an industry which in technology, research and development, promotional techniques and corporate structure, has exemplified modern food manufacturing.

The British market for breakfast cereals

Cereal-based health and baby foods have been manufactured in Britain since the late nineteenth century. In 1912 over sixty different brands of proprietary cereal foods are recorded, but few commanded a large following.[20] These included Falona (a mixture of wheat, oats, barley and bean flour) and John Bull (malted cereal and dried milk) as well as a number of North American products, imported either directly from the United States, or from Canada, through subsidiary companies of US firms. In 1893 the American Cereal Company, the forerunner of Quaker Oats, established an agency in London, and in 1899, a subsidiary company for the sale of rolled oats and later of Puffed Wheat, Muffets (a shredded wheat product) and Corn Flakes. In 1902 A.C. Fincken of Watford was appointed UK distributor for Force Wheat Flakes, manufactured by the Force Food Company of Canada. Shredded Wheat, made by the Shredded Wheat Company of Canada, was first imported by a firm of grocers in 1908, and shortly afterwards a UK subsidiary company was formed. Apart from 'Granose', there is no direct evidence that Kellogg products were imported before the Great War, but both Grape Nuts (made by the Postum Cereals Company) and Malta Vita (made by the Battle Creek Pure Food Co.) were recorded in 1912.[21]

The assault proper on the UK market, and the subsequent rapid broadening of the market base, began after 1920 when a number of leading American firms set up manufacturing plants in this country. In 1920, Quaker opened a small factory at Ware in Hertfordshire to produce Puffed Wheat and Puffed Rice, and in 1936, a large modern factory at Southall, Middlesex, which produced a wider range of products, to include, eventually, Quaker Oats. The Shredded Wheat Company started manufacturing at Welwyn Garden City in 1925. Kellogg began to import Corn Flakes and All Bran in 1922 and Rice Krispies in 1929, and in 1938 opened its first UK factory at Trafford Park, Manchester. The only major British enterprise, and that only partly so, because two of its three founders were South African while the product itself had Battle Creek and Seventh Day Adventist connections, was the African Cereal Company (now Weetabix Ltd), which set up a factory in a run-down flour mill at Burton Latimer, in Northamptonshire, in 1932. Indigenous enterprise was largely confined to the manufacture of porridge oats, which was carried on in numerous small mills up and down the country. Yet even here, as with other breakfast cereal foods, the industry was dominated by American firms. In the late 1930s Quaker controlled an estimated 62 per cent of the rolled oats market and Kellogg and Shredded Wheat over 80 per cent of the ready-cooked market.[22] [. . .]

The overall trends in the output and consumption of breakfast cereals are reasonably clear, but before 1950 difficult to quantify. Up to the Great War it may be assumed that the market for rolled oats expanded faster than that for RTE cereals, which were still a relatively high-cost speciality food. The RTE market grew dramatically between the wars, especially between 1936 and 1940 when Kellogg sales alone rose by nearly 80 per cent, and by the outbreak of war its value exceeded that of porridge oats.[23] In the Second World War the pattern of development was interrupted by a shortage of raw materials and by 'zoning schemes' which confined company sales to different parts of the country. Following the removal of wartime controls, the pre-war trend was resumed.

The total quantities of breakfast cereals entering the UK market can be calculated from production, from consumption, or directly, from retail sales. The evidence, however, is too patchy and unreliable for any one method to yield a continuous series for the whole of the period since 1900. [. . .]

With consumption we are on firmer ground in as much as there exist for the later 1930s a number of useful consumer surveys, and from 1940 onwards a more or less continuous series, compiled from official sources, of domestic food consumption and expenditure. Average consumption of RTE breakfast cereals in the UK in 1938 was estimated at 29 oz per head of the population.[24] Consumption in working-class households rose, despite the war, to 38–42 oz in 1942–45, and 62 oz in 1950.[25] The consumption of porridge oats increased (or recovered) during the war due to food shortages and reached a peak in 1947–8 following the failure of the potato harvest. [. . .]

The appeal of breakfast cereals lay in their palatability and convenience. The pattern of diffusion was, however, complex. Between the wars, ready-cooked breakfast cereals, oatmeal porridge and bacon and eggs competed directly with each other for pride of place on many middle and working-class breakfast tables. Consumer surveys, conducted in the 1930s, suggest a high price and income elasticity of demand for RTE cereals and bacon and eggs, and for both a significantly higher consumption in upper- than lower-income households.[26] In York, for example, a typical breakfast in the lowest income group was bread and dripping with a little bacon at weekends.[27] In southern Britain, where there was no long tradition of oatmeal-eating, porridge became popular in upper-class homes after 1870 and subsequently spread down the social scale. In Scotland, on the other hand, porridge was a traditional dish but its popularity was declining. Curious it was, that porridge had to cross the Atlantic and suffer a sea change before it became a utility food in England.[28] Before

1940 the pattern of change was, therefore, complicated. In the post-war period RTE cereals spread down and across the income scale, mainly at the expense of the hot breakfast. In 1936, according to the Crawford and Broadley survey, average expenditure on RTE cereals in income groups A and B was double that in Groups C and D, whereas in 1970 the income correlation, although still positive, was of a much smaller order.[29] The breakfast surveys of 1936, 1958, 1965 and 1967 show a progressive increase in the numbers of RTE eaters and suggest that by the late sixties ready-cooked cereals were the most popular weekday breakfast and that bacon and eggs were more usually reserved for weekends.[30] The decline of porridge-eating was a surrender to the convenience and for children especially, the frivolity of ready-cooked cereals. Rolled oats offered few opportunities for product innovation, so that technically, the different oatmeals competed with each other on speed of preparation. [. . .]

Advertising

Breakfast cereals have always been associated with intensive advertising and vigorous promotion which, in their method and general strategy, followed much the same lines as in North America. Historically, advertising expenditures have been high, normally between 10 and 20 per cent of net sales, but in the case of the Kellogg Company as high as 28 per cent in the mid-1930s.[31] It is estimated that in 1935 some £700,000, or about 18 per cent of manufacturers' net sales, was spent on proprietary breakfast cereals, as compared with 2.6 per cent on self-raising flour; 2.3 per cent on biscuits and 2.0 per cent on bread.[32] In 1938 cereal foods were said to account for about 10 per cent of total food advertising expenditure, more than four-fifths of which was taken by the four firms, Kellogg, Quaker, Shredded Wheat and Postum Cereals.[33] The emphasis of promotional effort shifted over time as the size and composition of the market changed and as new media became available. The strategy employed today is different from that in 1930 when appropriations were spent mainly on dealer aids and on advertisements in the national press, and different again than at the turn of the century, when, at the two extremes, the emphases were on tasteful advertisement in the better-class magazines, and outrageous stunts like the draping of advertising banners across the face of the white cliffs of Dover.[34] Since the 1920s advertising has been aimed mainly at children and at the more aggressive sections of the grocery trade, notably chain stores and supermarkets. [. . .]

Yet despite the thrust and aggressiveness of the industry, and the novelty of its products, a main feature of the post-war RTE breakfast cereals market has been the relative conservatism of the consumer, so much so that today over 70 per cent of sales by weight comprise the old pre-war favourites, Corn Flakes, Shredded Wheat, Weetabix, and Puffed Wheat. [. . .]

Conclusion

The difference between bread and breakfast cereals was only in part material and technological. Above all, it was a difference in concept, between the traditional and socially engineered product, that was apparent at each and every link in the chain from manufacturer to consumer. Bread was an 'inferior good' and the brown bread experience suggests that the potential of new products is weak where the benefits are intangible and the incentives are not deeply grounded, socially or dietetically, in the 'consumer revolution'. These differences are further reflected in the structure of the two industries. In the one, concentration has been primarily associated with marketing efficiency and the development of brand names; in the other with manufacturing efficiency and control of the retail trade. In breakfast cereals there has been continuous product development and innovation, whereas bread had relatively little scope or opportunity for diversification.[35] The basis of competition in breakfast cereals was not just one of price because in the mind of the consumer each offering was 'a different bargain' distinguished by way of trade mark, design formulation, taste, colour, and utility. In the late 1930s an estimated £700,000 was spent on advertising breakfast cereals compared with only £170,000 on bread.[36] [. . .]

Dietary change, like the standard of living itself, is a slippery concept. In 1929, J. Edgar Hoover analysed the essence of American prosperity as the stimulation and satisfaction of human wants to the point where 'the great mass of consumers will never for a moment know what it means to be content'.[37] This paper has suggested that role of the consumer in the growth of the modern food industry was rather more positive, and that changes in cereal-eating habits were the result of a more complex, and often unpredictable inter-reaction between supply and demand and between the market and the manufacturer.

Notes

1 Derived from C.H. Feinstein, *National income, expenditure and output of the United Kingdom 1855–1965*, Cambridge, 1972, Table 25, whose data were derived from A.R. Prest and A.A. Adams, *Consumers' Expenditure in the United Kingdom, 1900–1919*, Cambridge, 1954, and R. Stone and D.A. Rowe, *The Measurement of Consumers' Expenditure and Behaviour in the United Kingdom, 1920–1938*, 2 vols., Cambridge, 1953 and 1966. 'Food' includes all household consumption of foods and non-alcoholic beverages, plus garden and allotment produce, valued at farm prices, but excludes consumption in hotels and restaurants, which, if included, would have raised the post-war (cf. pre-war) food fraction by about 2 per cent. J.C. McKenzie, 'Food trends, the dynamics of accomplished change', in J. Yudkin & J.C. McKenzie (eds.), *Changing Food Habits*, 1964, p.41; *Household Food Consumption and Expenditure*, 1970 and 1971, Ministry of Agriculture, Fisheries and Food, 1973, (serial together with Supplement of food expenditure by working-class households, 1940–49, which is bound with 1950 report, hereafter referred to as NFS – National Food Survey).

2 For nineteenth-century trends in cereal-eating, see, E.J.T. Collins, 'Dietary change and cereal consumption in Britain in the nineteenth century' *Agricultural History Review*, vol. 23, 1975, pp. 97–115.

3 J. Yudkin & J.C. McKenzie (eds.), *Changing Food Habits*, 1964, pp. 40–43; P. Maunder, *The Bread Industry in the United Kingdom*, Dept. Ag. Econ. Univ. Nottingham, and Dept. of Soc. Science and Econ., Univ. Technology, Loughborough, n.d., pp.15–21.

4 Stone & Rowe, op. cit., p.26; NFS, 1966–70.

5 Census of Production, 1935, op. cit., p.46; *Report of the Census of Production for 1954*, Industry B, Table 5.
6 *Report of the Departmental Committee on Night Baking*, Cmd. 5525, 1937, para. 46; Maunder, op. cit., pp.21–7.
7 R. Evely & I.M.D. Little, *Concentration in British Industry*, Cambridge, 1960, p.254. An important enabling factor was the development of road transport after 1920 which brought the plant baker to the master baker's doorstep. For a summary of technical developments in the bakery industry see J.B. Jefferys, *Retail Trading in Great Britain 1850–1950*, Cambridge, 1954, pp.211–13.
8 Ibid., Ch. VII.
9 Evely & Little, op. cit., pp.257–8.
10 Collins, loc. cit.; NFS, 1970.
11 W. Tibbles, *Foods*, 1912, p.417.
12 For a comprehensive summary of the brown and white bread controversy see R. McCance & E. Widdowson, *Brown Breads and White*, 1956.
13 The foregoing information about the history of Hovis Limited has been kindly supplied by the firm of Rank Hovis McDougall Ltd. This is not intended as a comprehensive review of the activities of the firm but is concerned mainly with advertising and promotion. The more general history is summarised in *The Hovis Jubilee: a Brief Record of the Company's History between 1898–1948* (privately printed, London, 1948).
14 H.V. Edwards, 'Flour milling', in M.P. Fogarty (ed.), *Further Studies in Industrial Organisation*, 1948, p.46.
15 W. Crawford & H. Broadley, *The Peoples' Food*, 1938, pp.168–9.
16 NFS, 1950–1972.
17 The following survey of the historical development of the American breakfast cereals industry is based on Gerald Carson, *Cornflake Crusade,* 1959, and on information provided by the Kellogg Co., Battle Creek, Michigan, USA. The basic manufacturing techniques are explained in J.S. Remington, 'Breakfast cereal foods', *Food Manufacture*, Sept. 1935, pp.305–19, and more recent developments (in the USA) in *Studies of Organization and Competition in Grocery Manufacturing*, Nat. Committee on Food Marketing, Study No. 6, 1966, Ch. 9.
18 The transition from a health food with a limited appeal to a utility food with a popular appeal is perhaps the most important aspect of the historical development of breakfast cereals. Indicative of the new status of breakfast cereals were changes in nomenclature – 'Post Toasties' for 'Elijah's Manna', and the Kellogg Co. for the Sanitas Nut Food Company. Most breakfast cereals were originally marketed as 'natural', 'biologic' or, in the case of Grape Nuts, 'brain' foods. Post was a multi-millionaire by 1911 but W.K. Kellogg was the more successful entrepreneur; he had little sympathy with health reform and in 1911 bought out J.H. Kellogg's interest in the Battle Creek Toasted Corn Flake Co.
19 *Studies of Organization and Competition in Grocery Manufacturing*, op. cit. *The Structure of Food Manufacturing*, National Committee on Food Marketing, Technical Study No. 8, 1966, pp.11–13.
20 Tibbles, op. cit., p.443ff.
21 ibid.
22 Private communication: Quaker Oats Ltd.
23 *A Report on the Supply of Ready-cooked Breakfast Cereal Foods*, Monopolies Commission, 1973, p.5.
24 ibid., p.4.
25 NFS, 1942–50. The wartime estimates are crude and probably inaccurate as consumption has been derived from unit price and expenditure data.
26 e.g. Crawford & Broadley, op. cit.; J.B. Orr, *Food, Health and Income*, 1936; *Family Diet and Health in Pre-war Britain*, Carnegie Trust, Dunfermline, 1955.

27 B.S. Rowntree, *Poverty and Progress*, 1941, p.189.
28 Collins, loc. cit.; Crawford & Broadley, op. cit., p.39ff. In the USA oatmeal was first popularised as a breakfast food by a German immigrant, Ferdinand Schumacher, of Ohio, and the first advertisement appeared in a local newspaper in 1870.
29 ibid., pp.39ff, 166ff; NFS, 1970.
30 Crawford & Broadley, op. cit., p.39ff; G.C. Warren (ed.), *The Foods We Eat*, 1958, p.23ff; *Grocer*, 10 July 1965; *Grocer*, 8 July 1967.
31 *Breakfast Cereals*, Monopolies Commission report, op. cit., p. 5. Kellogg spent 13% in 1930–32, 20% in 1933, 27.7% in 1935 and 18.2% in 1939.
32 N. Kaldor & R. Silverman, *A Statistical Analysis of Advertising Expenditure and of the Revenue of the Press*, Cambridge, 1948, p.144; breakfast cereal ratios were exceeded by those of malted health foods (41%) and beef extracts and essences (22%), ibid., p. 144.
33 F.P. Bishop, *The Economics of Advertising*, 1944, pp.83, 87.
34 The crucial role of advertising in the growth of the modern food industry is a subject all by itself. The emphases were successively those of (a) posters and billboards, (b) national press, (c) television. Newspapers allowed display ads only after 1870 and this became a main source of revenue for national newspapers and popular magazines between the wars. Between 1910 and 1930 display advert space in eight London dailies more than doubled, and in four London dailies advertising volume (space *x* circulation) rose tenfold. (F.W. Taylor, *The Economics of Advertising*, 1934, p.212ff). In 1930 one leading RTE firm spent 35% of its advertising appropriation in newspapers and magazines, 10% on posters and 35% on dealer aids: ibid., p.228. For general reading see Taylor, op.cit., Bishop, op.cit., D. & G. Hindley, *Advertising in Victorian England, 1837–1901*, 1972; P. Hadley, *The History of Bovril Advertising*, 1974.
35 Figures for the UK are not available but in the USA it was estimated that in 1967 43 per cent of 'prepared breakfast foods' were new (i.e. post-1957) products. *Food Marketing and Economic Growth*, OECD, Paris, 1970, pp.36–7.
36 Bishop, op.cit., p.87.
37 Quoted Taylor, op.cit., p.47.

4.2 F.B. Smith, *Childbirth*, 1979

Despite progress in public health and medical science during the nineteenth century childbirth remained dangerous to women and infancy precarious for their children.

The anxious rituals associated with Victorian childbirth become more understandable when we begin to count the numbers of deaths. In 1847, the first year for which we have reasonably firm figures, over 3,200 childbirth deaths in England and Wales were reported to the Registrar-General, nearly 3,000 in 1861, nearly 4,000 in 1871, 4,200 in 1881, 4,400 in 1901, and 3,800 in 1903, the year when the incidence of maternal mortality began decisively, though still slowly, to decline. Measured against each 1,000 live births, these figures represent rates of 6 per 1,000 in 1847, 4.3 in 1861, 4.9 in 1871, 4.8 in 1881, 4.7 in 1891 to 4.8 in 1901, to 4.03 per 1,000 in 1903. The maternal death rate did not fall below 1 per 1,000 births until 1944–5.[1]

Source: F. B. Smith, *The People's Health 1830–1910*, London: Croom Helm, 1979, pp. 13–15, 28–31, 55–58.

This general rate can be divided into two categories, 'puerperal fever' and 'accidents', which indicate broadly the cause of death. Between 1847 and 1854 deaths reported as the consequence of 'puerperal fever' amounted to 1.72 per 1,000 and 'accidents' to 3.01 per 1,000.[2] During the period 1855–94 reported 'puerperal fever' deaths increased to 2.53 per 1,000, while 'accidents' declined steadily if slowly to 2.41 per 1,000. In 1903, the turning-point, 'sepsis' fell to 1.73 (from 2.13 in the preceding year) and 'Other Causes' accounted for 2.31.[3]

These rates are bad enough in terms of preventible deaths, broken families and orphaned children, but before we interpret them we must measure their limitations. They under-represent the numbers. Deaths in childbed were under-reported throughout the period: probably grossly so until the 1880s. Doctors, midwives, officials of lying-in hospitals and Poor Law guardians all had an interest in keeping the numbers down. Neither 'puerperal fever' nor 'accident' was defined for the purpose of registration throughout the century. 'Puerperal fever' meant at various times a single comprehensive disease covering diverse kinds of inflammation and symptoms, or congeries of differing forms of infection, or specific streptococcal infections, and also included excessive bleeding and paralysis of the limbs. 'Accident' was normally believed to cover ectopic pregnancies (the foetus developed outside the uterus in the uterine tubes) leading to abortion and sepsis, eclampsia (convulsions and coma associated with high blood pressure and fluid retention in late pregnancy) and 'exhaustion'. Only very rarely, after an unavoidable coroner's inquest, did 'accident' as a cause of death apparently include negligence by the *accoucheur*. Conveniently, then, until the 1880s when reporting procedures were tightened, women deceased in childbed who had shown symptoms of heart disease, dysentery or 'peritonitis' were entered under that particular heading, irrespective of whether they had also displayed more proximate causes of death, such as extensive infection or haemorrhage. Even after 1881, when the regulations were first improved, infections which set in ten days or more after delivery were also reported as causes of death independent of childbirth. Doctors, midwives and officials in some towns, notably Birmingham, simply refused to specify deaths in childbirth or from puerperal fever. In 1895, after an investigation of 4,000 deaths among women, the Registrar-General estimated that between 1885 and 1890 one-half of the deaths ascribed to 'metria' among women of child-bearing age, one-third of the 'blood poisoning, pyaemia etc.', and one-quarter of those reported as 'haemorrhage' were actually 'connected with childbirth', thereby indicating that, even allowing for yet worse earlier under-registration, the incidence of 'puerperal fever' held up throughout the later nineteenth century.[4]

Under-registration was probably greatest in regions of high maternal mortality and it would heighten what we already know of regional variations through the period. In general, as for example the decade 1881–90, maternal death rates for both 'puerperal fever' and 'accidents' were reported as lowest in London, where registration was effective, and the Home Counties, and highest in manufacturing, mining and rural districts, where registration was often poor. 'Puerperal fever' was the dominant ascribed cause of death in urban places, 'accidents' in rural and mining areas.

The general death rate of the south-west of England was below the national average, although its reported 'accident' rate was higher. The pattern indicated in Table 1 is related to social class and thereby to the kinds of attendance to be had, but not in any simple way. Within London, for example, wealthy parishes such as Hampstead and St George's, Hanover Square, had higher puerperal fever and general maternal mortality rates than wretched St Giles and St George in the East. Wrexham, a small manufacturing, semi-rural town, had the same reported incidence of puerperal fever deaths as Swansea and a higher maternal mortality.[5]

Table 1
Mean Annual Death Rate per 1000, 1881–90

	'Puerperal fever'	*'Accidents'*	*Total*
England and Wales	*2.59*	*2.13*	*4.73*
London	*2.44*	*1.53*	*3.91*
Manchester	*3.13*	*2.06*	*5.20*
North Wales – counties	*3.25*	*3.47*	*6.73*
South-west England – counties	*2.05*	*2.32*	*4.36*

Source: Dr Thursfield, *Lancet,* 7 January 1893, p. 21; *Parliamentary Papers*, 1884–5, vol. XVII, p. lxxiv.

[. . .]

In the towns women could use the out-patients' departments of the hospitals, or dispensaries, apply at a charity lying-in hospital, or simply wait until labour started, summon the doctor and have the baby delivered in the full knowledge that either he would not bother to ask for his money or that he would soon give up dunning the family. In country villages and towns it was common for local GPs to attend confinements and not charge for them. Dr John McVail, an experienced and level-headed investigator for the Royal Commission on the Poor Laws, concluded that in Edwardian times many country GPs provided about £50 of treatment a year, much of it midwifery, without seeking payment. This came about not only because the doctors were generous but also because country doctors still did not regularly keep case-records, especially of poor patients.[6]

If doctors liked charity because it conferred absolute control, suppliants were rendered more obsequious and anxious by its uncertainty. There is evidence that not all doctors were charitable at all times. In 1871 at Dymock, near Ledbury in Gloucestershire, a girl died while giving birth to illegitimate twins. The midwife engaged by the girl's father found the case difficult through the night and at 7.30 a.m. the girl's father went to fetch the parish Poor Law medical officer, Mr Cooke, who lived about three miles away. The father had no poor relief order and no means of paying privately. Cooke refused to come. At 5 pm the father persuaded Mr Wood of Ledbury, four miles away, to attend his girl. When Wood arrived the mother and twins were dead. Cooke told the coroner he had frequently attended cases without payment or a relief order. The *Lancet* concluded that:

> Mr Cooke is a gentleman who has been more than 40 years in practice, and who has reached a period of life at which he cannot

> be expected to be at the beck and call of every girl who has the misfortune to have an illegitimate child . . . The fact is, medical men are far too fearful of claptrap charges of inhumanity.[7]

At Mortlake in Surrey in 1897, Dr Macintosh was awakened at 2 am to attend a dying 3-day-old boy and his mother, described at the subsequent inquest as 'half-starved'. The child weighed 2 lb 2 oz at the post-mortem. After learning that the father could not pay, Macintosh went back to bed. Like Mr Cooke and numerous other doctors in similar situations at inquests, Macintosh told the coroner that he 'had attended 40 or 50 such cases and never received a penny', but he produced no corroborative details and as usual, the coroner did not press the matter.[8]

For admission to the books of a general hospital, a pregnant woman needed letters of recommendation from one or more governors of the hospital. This requirement, with exceptions such as the London Royal Free Hospital and the Edinburgh Royal Infirmary, lasted until the 1880s when hospital authorities began increasingly to dispense with 'lines'. To obtain the 'lines' the woman or her friends had to discover who the governors were and their addresses; some hospitals supplied lists, others did not; apparently the woman worked by local hearsay, with the consequence that some governors complained of being plagued and others chose to give out-of-town addresses. Whatever the outcome, the woman could not escape much walking and humble waiting. Evidence about what happened next is skimpy. The midwifery out-patients were uninteresting to hospital men at the time and historians of hospitals ignore them. Normally the woman had to go to the out-patients' department to show her 'lines' and give the porter her name and the address where she was to be confined; if she expected a troublesome labour she might seek to be examined by one of the junior house surgeons on duty. If the surgeon agreed, she had to wait an average of about four hours and, at some hospitals, such as the Glasgow Royal Infirmary until well into the 1880s at least, until all other out-patients were dealt with and the students departed, at 4 pm. The authorities thought it improper that students should observe the examinations of women.

If the woman went to one of the few hospitals to admit pregnant women, such as the London (after 1853), Guy's or the Dumfries and Galloway Infirmary, made her plea and proved an 'interesting' case, she might be admitted as an in-patient, up to one month before delivery.

Three things are clear about general hospital midwifery. First, the woman was almost invariably delivered at home; most hospitals excluded pregnant women, at least until the 1890s. If a woman admitted with illness was found also to be pregnant, she was promptly turned out. In 1844 a young woman was admitted with fever to the Westminster Hospital. Upon being washed and examined, she was discovered to be seven months pregnant, and was promptly sent home. Next day she died after a premature birth.[9] Second, the woman was delivered by the *accoucheur*-lecturer with students attending, or apparently more commonly, by a gaggle of students, sometimes with a midwife, unsupervised. Presumably the woman sent to the hospital for help when labour commenced. Third, the number of women delivered as out-patients was very large and the women were poor: in 1857 over 400 out-patients were confined by students from the London

Hospital; between 1854 and 1860 Guy's accepted 12,000 maternity cases from within a half-mile radius of the hospital, with a reported total of 36 deaths, or 1 in 333, a rate better than the national average. This reported rate is especially impressive because over 2,500 of the women were in their seventh or later confinements; one was up to her 22nd.[10] Dr Hicks, the physician-*accoucheur*, banned students in the early 1860s from attending confinements on the same day as they dissected. Guy's also dispensed with sponges in childbirth: flannels were used instead and then either burnt or washed in chloride of lime. The maternity department of St George's Hospital also had a good record: a death rate of 1 in every 284 deliveries between 1864 and 1876, and then 1 in 594 between 1869 and 1876.[11] In the early 1890s the West End Branch of the Glasgow Maternity Hospital was handling over 500 out-patient deliveries annually, and losing only five mothers a year.[12] The mother and her child were isolated from some of the infections endemic in hospitals and were delivered among friends and in familiar surroundings, however wretched. Their status, contemporaries remarked, ranged from the respectable artisan rank to the very poor. Most, given the usual hospital rules, had to claim to be married, at least.[. . .]

Stability and change in the maternal death rate

This survey of the evidence about maternity practices and maternal death rates suggests three general conclusions. First a woman and her infant did best if the birth was managed outside a hospital. Second, mother and child were safest, if the birth was a normal one, with a midwife; and if not, they were in grave danger. Midwives carried less infection with them, doctors brought instruments that in the last resort could save life. Third, there was no overwhelming displacement of midwives by *accoucheurs*, despite assertions to the contrary in recent historical-polemical writing. *Accoucheurs* probably extended their practice among the comfortable classes, while the increase in disposable income and the spread of the Poor Law system after 1834 probably enabled a growing number of poor women to dispense with neighbours and relatives and to have attention from experienced midwives. By the mid-1870s Dr Farr estimated that about 70 per cent of all births in England and Wales were managed by midwives.[13] In 1892 one informed observer guessed that midwives were handling about half the 870,000 labours a year in England and Wales.[14] The variations through Great Britain are enormous, and inexplicable except in local contexts which we have not yet explored. In 1869 the Obstetrical Society asked its Fellows to report on the proportion of attendance by midwives in their neighbourhoods over the past few years. The results run: for three villages – Fleggburgh, Norfolk, population 554, reported 30 per cent; Gringley-on-the-Hill, Nottinghamshire, population 874, 46 per cent; Bromyard, Herefordshire, population 1,300, 90 per cent. The average for the reports from 'small manufacturing towns' at 6,000–10,000 populations ran at about 5–10 per cent; Lewes was 'nil'; Long Sutton in Lincolnshire was 26 per cent; Altrincham in Cheshire, 53 per cent, 'Large manufacturing towns', Glasgow, for example, ran at about 75 per cent, and Edinburgh almost none; Coventry 90 per cent. In Wakefield, all the

Irish were delivered by midwives, all the English by doctors. The parishes of the East End of London varied between 30 and 50 per cent; the West End was 'very slight' at under 2 per cent. The availability of students from medical schools made a difference, because women preferred them, at least in Edinburgh and the East End. There was always a crisis and a rush for midwives during the hospital vacations. The ratio of confinements managed by midwives was still varied through the country in 1909. Over all, it was said to be about 50 per cent, but in Newcastle it was only 11.2 per cent, in the London County Council area 25 per cent, in the West Riding 35 per cent, Liverpool 52 per cent, Manchester 60.9 per cent, Gloucester 83.6 per cent, and 93 per cent in St Helens. The availability of local charity funds must also have shaped the pattern of attendances, especially in richly endowed towns with a low birth rate, such as Exeter, or towns under local Acts outside the Poor Law, such as Coventry and Brighton.[15] This might explain the 100 per cent doctor attendance in Lewes. But as the Wakefield figures show, there are no simple answers. Were the Irish boycotting Protestant doctors? Were the doctors boycotting the Irish? At first sight it could be a case of Catholic women preferring to be delivered by women, but Irish women at home and elsewhere in Britain happily went to doctors for their confinements. Perhaps the Irish came within the rubric of a local Irish midwives' charity? But these speculations only illustrate how little we really know, and why sweeping assertions about the midwives/doctors question are ill founded. What is clear is that, whether delivered by a doctor or a midwife, the mother had to be strong, and lucky, to survive.

The stability of the maternal death rate throughout the century suggests that antisepsis and improvements in obstetrics made little difference. Indeed, as I have suggested, the increased resort to instruments might have maintained the rate during the last third of the century and into the 1930s. The death rate for 'accidents' held up until 1941–2. The servant-employing classes who retained expensive *accoucheurs* bought greater peace of mind, but no real lessening of their vulnerability. Fifty years of expensive education in midwifery ought to have made doctors less rough and more resourceful in a crisis, but measured against the death rate it does not appear to have done so. The spread of trained midwives must have helped lower the rate in the twentieth century, but again one would have expected their impact to have been greater than it appears to have been. This was partly because the trained midwives were not getting to the women who needed them most.

In Rotherham, for instance, where midwives were hand-picked by the health authorities and especially encouraged, 25 per cent of births in 1907 and 1908 were still 'unattended' and 25 per cent were attended by 'old unqualified women'. 'Some women', Dr Robinson, the MOH, alleged, 'go unattended or employ "unqualified women" because they have a reputation for having a large amount of what we call in Yorkshire "churchyard luck" [i.e. a high proportion of the infants they delivered were 'stillborn'].'[16]

The death rate from toxaemia also remained almost unchanged until it was halved between 1941 and 1945. Clearly the spread of the new penicillin drugs and the improved nutrition of expectant mothers during the war made the great breakthrough.[17]

The clue to the first transition in 1903–7 might lie in the decreased number of pregnancies past the eighth (and a consequent reduction in the number of women bearing children at the more dangerous end over 30 of the child-bearing age range). Maternal deaths at the eighth confinement, following Australian experience during the period 1892–1900, probably became statistically more likely than at primiparae.[18] There are no comparable data available for Britain, but there is no reason why the Australian situation should not apply. This reduction ultimately depends on greater recourse to contraception and its greater effectiveness.[19] We have no detailed information about the effectiveness of contraceptive practices and the sale of devices in this period,[20] but it is worth recalling that the decline in the English birth rate begins in 1877, in the same year as the rumpus in the daily press, the Church and the judiciary about the spread of information about family limitation. If family limitation is the key to the breakthrough, it is also worth recalling, in the light of the implicit medical cost-benefit sum, that medical doctors, with a few notable exceptions, determinedly fought the spread of contraception (at least in their public utterances) to mothers who practised or at least approved family limitation, and were themselves in the midst of their child-bearing years. And from what we know of present custom, it is a likely hypothesis that mothers encouraged daughters in family limitation and passed on contraceptive knowledge. Over all, it is worth emphasising that the maternal mortality is lower than we might have expected, given the poor nutrition and fatigue of overworked pregnant women, the risks of infection, the want of skill in the attendants and other hazards. But, as in modern Indonesia or Mexico, human reproductive processes outmatch poverty, ignorance and anaemia.[21] While 3,200 died in childbirth in England and Wales in 1847, over 450,000 lived (i.e. 1 death in 140); 3,000 died in 1861 and over 700,000 lived (1 death in 233); 4,400 died in 1891, and 860,000 lived (1 death in 198).[22] Here, the bias to mortality in our evidence is vividly exposed. One is left with an impression, both awesome and tantalisingly untraceable, of prodigious private stoicisms and strengths.

Notes

1 Dr William Farr, *Lancet* (hereafter *L*), 11 Aug. 1860, p. 138; *Parliamentary Papers* (hereafter *PP*), 1884–5, vol. XVII, p. lxxiii; Dr Thursfield, *L*, 7 Jan. 1893, p. 21; Janet M. Campbell, *Maternal Mortality* (London, 1924), p. 3; J.W.B. Douglas and G. Rowntree, *Maternity in Great Britain* (Oxford, 1948), p. 36.
2 *L*, 30 June 1866, p. 721.
3 Dr Farr, *L*, 26 Oct. 1872, p. 606; *PP*, 1895, vol. XXIII, part I, p. xxii; *PP*, 1904, vol. XIV, p. lxvi.
4 *PP*, 1895, vol. XXIII, part I, p. xxii; Dr W. Williams, *L*, 2 Jan. 1897, p. 50.
5 *PP*, 1895, vol. XXIII, part I, p. 685.
6 *PP*, 1909, vol. XXXXII, pp. 124–5.
7 *L*, 1 Apr. 1871, p. 454.
8 *L*, 30 Jan. 1897, p. 330.
9 John Langdon-Davies, *Westminster Hospital: two centuries of voluntary service, 1719–1941* (London, 1952), p. 61.
10 J.C. Steele, 'Numerical Analysis of the Patients treated in Guy's Hospital . . . 1854–61', *Journal of the Statistical Society*, vol. XXIV (1861), pp. 388–99.

11 *L*, 7 July 1877, p. 22.
12 Robert Jardine, 'Notes of 1028 Cases . . .', *Glasgow Medical Journal*, vol. XXXIX (1893), pp. 32–5.
13 Jean Donnison, *Midwives And Medical Men* (London, 1977), p. 77.
14 Dr J.H. Aveling, Registration, *PP*, 1892, vol. XIV, Qs. 251–3.
15 'Report on Midwives Act', *PP*, 1909, vol. XXXIII, p. 4.
16 Ibid., Qs. 2458–9, 2597.
17 Richard M. Titmuss, *Problems of Social Policy* (rev. ed., London, 1976), pp. 512–23, 535–7.
18 Campbell, *Maternal Mortality*, pp. 6–10; 'Maternal Mortality in Connection with Childbearing', *PP*, 1914–16, vol. XXVI.
19 See N.D. Hicks, *This Sin and Scandal. The Australian Population Debate 1891–1911* (Canberra, 1978).
20 J. Peel, 'The Manufacture and Retailing of Contraceptives in England', *Population Studies*, vol. XVII (1963–4).
21 Alan D. Berg, *The Nutrition Factor; its role in national development* (Washington D.C., 1973), pp. 15–16.
22 Calculated from *PP*, 1884–5, vol. XVII, p. lxxiii, *L*, 7 Jan. 1893, p. 21 and B.R. Mitchell and Phyllis Deane, *Abstract of British Historical Statistics* (Cambridge, 1962), pp. 29–31.

4.3 Anthony S. Wohl, *The Canker of Industrial Diseases*, 1984

It is dangerous to assume that death rates issued by the Registrar-General and other official sources for various occupations accurately reflect working conditions or the incidence of industrial disease, for a wide variety of influences were at work, among them domestic hygiene and housing conditions, public sewerage and water, diet, and general living standards. Further complicating the issue was the elimination of the weak from certain arduous trades. As Dr Ogle, at the Office of the Registrar-General, wrote, many industries 'are in fact carried on by a body of comparatively picked men; stronger in the beginning and maintained at a high level by the continual drafting out of those whose strength falls below the mark.'[1] Greenhow, Arlidge, and others interested in public health and occupational diseases were careful to point out how difficult it was to draw up any simple causal connection between certain hazards on one hand and the death rate among operatives on the other. It was argued, for example, that the high death rates prevailing among cutlery- and file-workers and the low rates among miners might in part be explained by the notoriously high incidence of drunkenness among the former and the general sobriety of the latter. The following tables, covering the years 1890–92, and compiled by Dr Ogle from some 500,000 returns, should be read then with due caution. The figures relate only to deaths of male workers between the ages of forty-five and fifty-five, the period in which, we may assume, the ravages of slow disease began to take their toll.

Source: Anthony S. Wohl, *Endangered Lives: Public Health in Victorian Britain*, London: J. M. Dent & Sons Ltd, 1984, pp. 279–84.

Table 1

Death rates for males, between 45 and 55 years of age, per thousand living[2]

Occupation	*Death rate*
Inn-keepers	44.48
Pottery, Earthenware	42.97
Dock, Wharf labourers	40.71
File makers	40.06
Lead workers	37.62
Costermongers	37.08
Cutlery, Scissors	35.60
Tin miners	33.20
Glass	32.14
General labourers (London)	31.94
Chimney Sweeps	31.43
Carmen and Carriers	28.01
Brass	26.05
Musicians	26.01
Cotton (Lancashire)	25.11
Coal miners (S. Wales)	24.27
Butchers	22.65
Bricklayers	22.04
Tailors	21.98
National Average	21.37
Medical Men (Surgeons, G.Ps)	21.04
Blacksmiths	20.74
Wool (West Riding)	20.58
Fishmongers	20.13
Tin-plate	20.08
Coal miners (national average)	19.42
Paper	18.84
Fishermen	18.61
Barristers	17.72
Railway porters	16.98
Coal miners (Northumberland & Durham)	16.35
Railway engine drivers	16.09
Domestic servants	15.85
Grocers	14.34
School-teachers	14.31
Farm labourers	13.56
Hosiery	12.15
Clergyman	10.52
Farmers	10.16
Artisans, Mechanics	8.83

Some brief word of explanation is needed for some of these figures. Inn-keepers at the head of the list as the most unhealthy occupational group might cause some bewilderment until it is realized that the Registrar-General included alcoholism and disease of the liver in his figures. Apparently many inn-keepers were their own best customers, for they had a death rate from alcoholism some seven times greater than the national average and some six times higher for both gout and liver diseases! Ten times as many London publicans died from alcoholism as the national average for all employed males. Pottery and earthenware workers suffered greatly from respiratory diseases affecting the heart and lungs, as did metal workers, and file-makers had a death rate almost double the national average for gout, phthisis, respiratory diseases, nervous diseases, and urinary infections. Lead-workers had a death rate almost three times higher than the national average for nervous diseases, two-and-a-half times higher for deaths from digestive diseases, and some four times higher for deaths from urinary diseases.[3] The high mortality of dock and wharf labourers should also be noted, indicating, perhaps, their general standards of living and the hazards of working in an undernourished state outdoors in London's climate: among the causes for their high mortality were consumption, pneumonia, and other respiratory diseases. Interestingly, as a group, general labourers, who, unlike factory workers, received no protection from general industrial sanitary and health codes, had a higher mortality rate at each age group than the factory operatives.[4]

Table 2

Occupation	*Deaths from Diseases of the Respiratory System*
Agricultural workers	100
Coal Miners	133
Wool Manufacturers	202
Cotton Manufacturers	244
Lead Workers	247
Chimney Sweeps	249
Brass Workers	250
Zinc Workers	266
Copper Miners	307
Copper Workers	317
Lead Miners	319
File Makers	373
Tin Miners	400
Cutlery, Scissor Makers	407
Potters, Earthenware Workers	453

That many industrial occupations were still hazardous at the end of the century is perhaps better illustrated by the table above, taken from the same comprehensive study conducted by the Office of the Registrar-General.[5] The statistics are for male workers only: for comparative purposes a base figure of 100 is used for agricultural workers. Once again there is a danger of too simplistic a connection between the trade and the incidence of death; yet deaths from diseases of the respiratory system do suggest the degree of atmospheric purity of the workshop or factory.

Appalling as is this wide range of mortality according to occupation, it tells only part of the story and does not convey the extent of the suffering brought on by slow, but not fatal illnesses, or the illnesses which other members of the family suffered as a result of privation when the principal bread-winners were forced from their jobs by occupational disease. Local authorities pointed out that the victims of industrial diseases and their families often dropped into the ranks of paupers. The observation (which we noted earlier in this chapter) of Dr White at the end of the century on the file-cutters of Sheffield applies with equal force to countless trades throughout the country. In industry after industry workers who had contracted some form of occupational disease did, indeed, 'too frequently become a burden on the community' or on their own families. Thus 'the canker of industrial diseases' must be numbered among the causes of pauperism and family destitution.[6]

In 1904, some three years after the various factory and workshop acts were consolidated, the Inter-Departmental Committee on Physical Deterioration maintained that the new act had been neglected in the Potteries, where it was most needed, and that there was still inadequate control over children in poor health entering workshops. Workshops were still largely beyond the arm of the factory laws, and in 1901 there were far more workshops than factories – 143,065 compared to 97,845.[7] In Leicester, when the small, old hosiery workshops gave way to factories in the 1870s there was a much more satisfactory control of working conditions, with a corresponding improvement in the health of the workers.[8] When, at the end of the century, a House of Lords Select Committee followed up the investigations of a decade of social reformers and went into the sweatshops and small workshops of East London, they were appalled at the lack of ventilation and the disregard for the health of the girls, and shocked at the ease with which the sanitary laws were circumvented. Inspectors could be spotted a mile off, workgirls were sent home (to avoid prosecution for overcrowding) and windows, normally shut tight, were hastily opened at the first sight of an inspector. Health regulations in the little sweatshops that were characteristic of the cheap clothing industry remained a farce throughout the late Victorian period. It was necessary for inspectors to prove that the labour performed in the home or workshop constituted 'the whole or principal means of living of the family' for the domestic setting to come under the provisions of the laws governing workshops. This requirement, combined with the reluctance of the authorities to invade what appeared on the surface to be a home, put sweatshops and much domestic industry beyond reach of the law. The 1891 Factory Act, following the Report of the House of Lords Select Committee on the Sweating System, permitted inspection of workshops without a writ or warrant and marked the beginnings of somewhat closer control of workshops.[9]

The Inter-Departmental Committee on Physical Deterioration discovered, much to its dismay, that the laws protecting young children were also often ignored. Although the new Factory Act of 1901 required the physical examination of all young persons applying for factory work, there was no such provision for youngsters in workshops, and although it did apply to mines, it was often ignored. 'It does not matter to the managers', one witness told the Committee, 'whether they [young miners] are scrofulus, rickety, phthisical, or anything else – they get them into the pit.'[10] In 1904 there were some 2,000 certifying surgeons to examine young workers entering the factories, but their main contribution lay in the future, and at the end of Victoria's reign there were still far too many young men and women who entered a life of factory work but whose physical condition was not up to the labour expected of them, or sufficiently sound to withstand the heat, humidity, overcrowding or contaminated air still encountered in many factories.[11] One must unfortunately conclude that, despite the many improvements in the final third of the century, the general observation of two early historians of the factory reform movement is pertinent for the control of industrial diseases and the provision of healthier working surroundings:

> We cannot adopt the enthusiastic tone of some who have written of the English Factory Acts as if the conditions of labour had thereby been completely transformed and humanised. On the contrary, closer study reveals the fact that an extraordinary timidity has beset all our efforts in the matter.[12]

Had Britain's industrial workers been able to leave the factories and workshops at the end of the day and return to healthy homes, perhaps their working conditions would not have exacted too heavy a toll upon their health. What these workers needed after a day's labour was rest and a modicum of domestic comfort. To sustain good health and to recuperate they required fresh air, good ventilation, sound sanitation, running water, and a basically sound diet. As we have seen, these last three requirements were all too often the exception rather than the rule, and certainly not widely available until the final quarter of the century. [. . .]

Notes

1 'Forty-fifth Annual Report of the Registrar-General, for 1897', p. 23, quoted in J.T. Arlidge, *The Hygiene, Diseases and Mortality of Occupations* (1892), p. 15.

2 Adapted from *PP. [Parliamentary Papers]* XXI (1897), 'Supplement to the Fifty-fifth Annual Report of the Registrar-General, Pt. 11 (1897), "Letter to the Registrar-General . . ."', pp. cxix ff.

3 For an indictment of the lives and habits of the publicans see Arlidge, *The Hygiene, Diseases,* pp. 152–5.

4 *PP.* XXI (1897), 'Supplement to the Fifty-fifth Annual Report of the Registrar-General, Pt. 11 (1897), "Letter to the Registrar-General . . ."', pp. lv, lvii, lxxv.

5 Ibid, p. xxi. These figures should be compared with those in *PP.* XXVIII (1864), 'Sixth ARMOHPC, for 1863', p. 24, in which Simon showed that the female death rate from phthisis and other lung diseases at ages 15–25 was

two-and-a-half times higher in Macclesfield and Leek (where many women worked in the silk factories) than among female workers as a whole.

6 *PP.* XII (1899), 'Third Interim Report of the Departmental Committee appointed to Inquire and Report upon Certain Dangerous Trades', p. 7.

7 *PP.* XXXII (1904), 'Report of the Inter-Departmental Committee on Physical Deterioration, I. Report', p. 30.

8 J. Simmons, *Life in Victorian Leicester* (Leicester, 1971), p. 77.

9 Even the 1901 Factory Act made domestic industry comply with the various factory and workshop acts only if the activities carried on there were certified as dangerous, see B.L. Hutchins and A. Harrison, *A History of Factory Legislation* (New York, 1903), pp. 208, 238–9.

10 *PP.* XXXII (1904), 'Inter-Departmental Committee on Physical Deterioration, I. Report', p. 29.

11 Ibid., see especially the evidence of Miss Adelaide Mary Anderson, the Principal Lady Inspector of Factories, pp. 63ff.

12 Hutchins and Harrison, *A History of Factory Legislation*, p. 253.

4.4 Eric Ashby and Mary Anderson, *The 'Pokeable, Companionable Fire'*, 1981

The lobby to dispel domestic smoke

[Despite legislation in 1858, 1866 and 1875 to curb industrial smoke] no one had ventured to invoke state interference with smoke from the homes of British citizens. In 1880 it was reckoned that there were 597,285 homes in the metropolis, with 3,583,000 fireplaces, and in 1881 seventy-two houses a day were being added.[1] So by the year 1880, when we take up again the campaign to abate smoke, the emphasis had shifted from industrial smoke in the north, conveniently remote from parliament, to domestic smoke in London, persistently visible from the windows of the House of Commons.

Even in these days of mass media it may be a book, like Rachel Carson's *Silent Spring*, that ignites the public conscience over some environmental hazard. Just such a book was *London Fogs* by the Hon. F.A.R. Russell (Lord John Russell's son), issued toward the end of 1880 and widely quoted for a decade afterwards.[2] All Londoners had experienced fogs but it was Russell's diagnosis that gave meaning to their experience. He noticed something that is still evident in the way people subjectively assess risks. If quite a small number of people are killed in one event in one place (say seven in a railway accident) it makes a much greater impact on the public than a far greater number of deaths dispersed in time and place (such as the 7,000 or so persons killed in Britain every year in road accidents). Russell made a similar observation for deaths due to London fogs. A London fog, he wrote, 'performed its work slowly, made no unseemly disturbance, and took care not to demand its hecatombs very suddenly and dramatically'.

In December 1873 there occurred one of the thickest and most persistent fogs of the century. *The Times* for 12 December had a leading article about it, declaring how it paralysed trade, inflicted a loss of time and money, and suffocated cattle on show at Islington. There was no mention in *The*

Source: Eric Ashby and Mary Anderson, *The Politics of Clean Air*, Oxford: Clarendon Press, 1981, pp. 55–59, 64.

Times leader of hazard to human health; on the contrary, the poor beasts at Islington 'may be considered to have been the chief sufferers'. It was Russell's important contribution to demonstrate that cattle were not the chief sufferers. From the Registrar General's statistics he calculated that during the week of the 1873 fog deaths in London exceeded the average by 700, of which 500 could be attributed to the fog. Late in January 1880 there was another visitation of fog; in the three weeks ending 14 February 1880 there were 2,994 additional deaths in London of which some 2,000 were attributed to the fog. These figures demonstrated that fog, in its undramatic way, was as ruthless a killer as cholera was. 'Such are the results', wrote Russell, 'more fatal than the slaughter of many a great battle, of a want of carefulness in preventing smoke in our domestic fires.' That domestic fires were the chief cause, Russell deduced from the fact that some of the worst fogs occurred on Sundays, and Christmas Day 1879 was spent in 'nocturnal darkness' although no factories were operating.

With a thoroughness not matched again until the 1950s, Russell speculated on the cost of damage done by fogs in London. Materials, from fabrics to stonework, suffered. 'The sitting figures . . . on the north side of Burlington House might, but for their European garb, be taken for Zulus.' The energy London received from the sun, he reckoned, was only about two-thirds of that amount falling on country areas. People had to live in the suburbs to escape fogs and this obliged them to spend money commuting to London every day. Property values in London were depressed by fogs. In a lecture he gave eight years after his book was published, Russell calculated that the annual cost of damage due to fogs was of the order of £5.2 millions, equal to the wages of 100 000 labourers.[3] It is ironic that two great comforts of Victorian life – the coal fire and the water closet – were both indicted for murder: the one through fogs which exacerbated respiratory diseases; the other which carried typhoid and cholera through sewers into drinking water.

Russell's suggestions for abating smoke from domestic fires are in a surprisingly modern key: encourage the use of coke and stone coal (anthracite) by putting a bounty of two shillings a ton on stone coal and a tax on bituminous coals; tax all new fireplaces of uneconomical design; tax cooking ranges not adapted to consume their own smoke; organize hire-purchase schemes so that people can buy well designed stoves; let the Metropolitan Board or some other public authority purchase the gasworks, to supply cheap gas and coke.

Between the 1850s, when smoke in London prompted Palmerston to act, and the 1880s, when Russell's book was published, the number of days of fog per annum in London had almost trebled. Pollution from industry was undoubtedly alleviated to some extent over this period by the laws controlling smoke and by the Alkali Acts. Pollution from the domestic hearth had got steadily worse; the quantity of coal brought to London more than doubled between 1862 and 1882.[4] Political action to abate industrial discharges had proved difficult enough; political action to abate pollution from private homes posed social problems of quite alarming dimensions. In an attempt to tackle these problems the smoke-abatement lobby came into being and played an important part through the rest of the nineteenth century.

There had, of course, been smoke-abatement lobbies before. But in the 1880s the movement was organized in a much more professional way. It started in October 1880, when a Fog and Smoke Committee, under the auspices of the National Health and Kyrle Societies, held its first meeting. The Kyrle Society was devoted to the creation of open spaces in London, part of the compassionate movement associated with the names of Miranda and Octavia Hill to bring beauty into the lives of the poor. The National Health Society's president was Ernest Hart, a surgeon and propagandist for sanitary reform. He became chairman of the Fog and Smoke Committee, recruited to its membership some very distinguished people, and guided its activities with statesmanlike skill.

Before bringing anything to parliament the committee decided to encircle the problem from four directions. It appointed a group to study the state-of-the-art of fuel technology; it asked a lawyer to report on the state of the law on smoke prevention; it enlisted public interest by organizing an exhibition of equipment and design for the abatement of smoke; and it invited some factory owners to confer with them as to how smoke could be abated with the minimum of interference with the manufacturer's interests. Its whole approach was a model in the art of lobbying.

After securing what was, for those days, good press coverage (even its meetings were reported in *The Times*) it began to apply tentative prods of pressure. It sent a circular to the vestries and district boards in the metropolis, reminding them that local authorities already had power to deal with nuisances caused by any 'mill, factory, dye house, brewery, bakehouse or gasworks' which did not, so far as practicable, consume their smoke. Someone saw to it that this circular was noticed in *The Times*.[5] The Committee then persuaded the Lord Mayor of London to allow it to have a meeting in the Mansion House. Again, press coverage was good. *The Times* announced the list of speakers (which included the president of the Royal Society and the First Commissioner of Works) and on the Monday after the meeting (10 January 1881) carried a full account of the occasion.[6] It must have been a heartening event. Notables like the Earl of Aberdeen and the Dean of Westminster were present; there were letters of support from Lords Derby and Aberdare; and Mrs Gladstone, who had evidently been invited, sent belated apologies from Downing Street and added: 'The nuisance of smoke is, indeed, horrible. Mr Gladstone and I often speak of it, and I may express our own feelings most truly in saying how thankful we should be if wise steps be taken so as to bring the matter forward in a practical manner.' If allowances are made for the circumlocutory caution which a prime minister's wife has to display, this was quite an encouraging message. Again, the committee saw to it that this message appeared in *The Times*.[7]

The First Commissioner of Works came to the Mansion House meeting briefed with the government's policy; his speech was a bland mixture of pious hopes and bureaucratic caution. He hoped the time would not be far distant when they would have restored 'the atmosphere of London to its early purity, the blossom to our London roses, and the bloom to the cheeks of our London children'; but he then kept his hopes at a politically safe distance by saying that he would 'deprecate any hasty attempts to legislate', and over domestic smoke he held out no expectation of legal enforcement. It would not be wise, he said, 'to attempt to interfere by any legislation. They must rather trust to persuasion and example and inducements . . . he did not see the means of persuading the enormous mass of householders to use the smokeless coal unless it could be distinctly proved to them that there would be economy in the change.'[8]

There was a hint to the committee not to rush its fences. It contented itself with a statesmanlike decision that the investigation and testing of various smokeless fuels and appliances 'should precede any application for amendment of the existing Smoke Acts, or for new legislation in regard to smoke from dwelling houses'.

It soon became apparent to the Smoke and Fog Committee that the intractable problems in the cure of domestic smoke were not in technology or economics: they were in sociology. There were already stoves on the market which would burn coke or smokeless anthracite coal. These fuels cost at that time no more than ordinary coal. But the householder, eager enough by now to suppress smoke from factory chimneys, was unwilling to forego his own cheerful fireplace for the silent and colourless emanations from a closed stove; nor, by all accounts, were servants anxious to exchange a daily chore of brushing up cinders and laying new fires for the cleaner job of tending an anthracite stove. (At one meeting of the Fog and Smoke Committee Mr Owen Thomas offered 'to send a couple of Welsh maids to town to show London maids how to burn smokeless coal in stoves suitable for domestic purposes'.)[9] So the top priority in the Committee's campaign was to overcome the inertia of householders.

This was the purpose of the exhibition which opened in South Kensington on 30 November 1881. It was launched with an impressive ceremony in the Albert Hall. Public interest was such that the assembled company, which included Princess Louise and the Marquis of Lorne, was 'numerous enough to fill the entire area of the hall'. The exhibits included a great variety of improved grates, stoves, and furnaces for domestic and industrial use, and an assortment of fuels. More than 100 000 people visited the exhibition, including the Prince of Wales and the Empress Eugénie. Special arrangements were made to bring parties from distant parts of London and from the provinces. When the show closed in February 1882 it was reopened in Manchester. There was good coverage in the press. The most fortunate and timely support came from the environment itself: on 18 January and again on 3–4 February there were awful fogs. London traffic came to a standstill; cases in the police courts had to be deferred because witnesses lost their way; theatres were practically deserted and those plucky enough to venture out to see a play found fog obscuring their view even inside the theatre.[10]

The exhibition was not merely a showpiece; the exhibits were subjected to quite sophisticated tests. The heating power per pound of coal consumed and smoke produced was worked out. Measurements were taken of the amount of unburnt carbon, the hydrocarbons, and the weight of solid matter intercepted on filters. On these and other criteria awards were made for the most efficient stoves in each class.[11]

All this publicity implanted the idea that smoke was an avoidable nuisance which ingenious designers were working vigorously to dispel. There was tangible evidence that some people were responding to the exhibition's message; one firm which displayed a cheap stove at South Kensington sold

14,000 of them in the succeeding two years.[12] But expectations that there was an acceptable cure for domestic smoke were disappointed. There was no doubt that closed stoves burning anthracite could dispel the nuisance, but there was also no doubt that Englishmen did not like closed stoves. The smoke-abatement lobby had to consolidate itself in order to have the stamina for a long campaign. This it did in 1882, by translating the informal Fog and Smoke Committee into an incorporated National Smoke Abatement Institution (NSAI). The lobby had by now mobilized very impressive support: the Duke of Westminster as president; other members of the nobility; distinguished professional men including the president of the Royal College of Surgeons; and men who had already made history in the applications of science to social reform: Edwin Chadwick and Lyon Playfair.

At this stage the lobby might have been tempted to press for some action in parliament. It wisely resisted any such temptation, for the case for bringing domestic smoke under the law was still weak. The man in the street was still not willing to barter his open fire for cleaner air and fewer fogs. [. . .] Even the scientific journal *Nature*, quoting the opinion of a distinguished sanitary engineer, Sir Frederick Bramwell, agreed: 'We are strongly of the same opinion . . . that we must have an "open, pokeable, companionable fire".'[13]

The obstacle to smoke abatement was not primarily technological: it was sociological. Sociological obstacles to progress are every bit as intractable as are technological obstacles, and we still know far too little about how to tackle them. It would be correct to say that in the 1880s there was already a best *known* means to abate smoke from domestic hearths. But it was not a best *practicable* means, for the problem of how to persuade people to do without visible flames had not been solved.

It was over the surveillance of a law covering domestic smoke that the most emotive arguments were used. It is not necessary to enter a house in order to see smoke coming out of the chimney. But much was made about 'an army of police watchers' and the intrusion of inspectors into private homes. A more serious question was how the law was to be administered in the Courts. The Bills put before parliament gave no guidance about the interpretation of their technical provisions. For example, all fires smoke when they are first lit: for how many minutes was such smoke to be tolerated? And how were the escape clauses – the defence that 'reasonable measures' had been taken to prevent smoke – to be operated? Who was to be held responsible for infractions: the ground landlord (who might have leased the house for fifty years), or the tenant (who might be coming to the end of an annual tenancy)?

The reformer's path was littered with obstructions such as these, some of them substantial, others mere quibbles. Beneath the prevarication there was, of course, a much deeper motive for resistance against laws to abate domestic smoke. Under common law most domestic smoke could not be proved to be harming anyone or anything; to control it would be to make something illegal which had been legal for centuries. This was a forfeit of liberty not to be made lightly.

There is an object lesson in the episode we have described in this chapter. It is the need to understand not just the technology of an environmental problem, but the mechanics of slow maturation of public opinion about it. *The Times*, looking back on the decade of the 1880s, put it well:

> We must . . . endeavour to stir up public opinion to the point of action by implanting a conviction in the public mind that civilisation itself is retarded by the toleration of nuisances that can be removed and of dirt that never ought to have been created . . . we are not without hopes that the day is not far distant when public opinion will insist that London must somehow be relieved of its canopy of gloom and its hideous vesture of dirt.[14]

Notes

1 *The Times*, 5 Nov. 1880, 8e; 13 Nov. 1883, 2f.
2 Russell, F.A.R., *London fogs*. Stanford, London (1880).
3 Russell, F.A.R., *Smoke in relation to fogs in London*. National Smoke Abatement Offices, London [1889].
4 *The Times*, 13 Mar. 1882, 12b.
5 Ibid., 18 Nov. 1880, 6d.
6 Ibid., 10 Jan. 1881, 4f.
7 Ibid., 21 Feb. 1881, 8a.
8 Note 6.
9 *The Times*, 5 Nov. 1880, 8e.
10 Ibid., 19 Jan. 1882, 10f; 6 Feb. 1882, 7e.
11 Smoke Abatement Committee, *Report . . . 1882, with reports of the Exhibition at South Kensington . . .* Smith, Elder & Co., London (1893).
12 *The Times*, 18 July 1883, 6f.
13 'Smoke abatement exhibition', *Nature* 25, 219 (1882).
14 *The Times*, 23 May 1890, 9d.

4.5 Nicholas Bullock and James Read, *The Movement for Housing Reform in Germany and France*, 1985

Our study starts from the premise that the development of the movement for housing reform in England, in Germany and in France had much in common. This was not only a natural consequence of the variety of contacts, both personal and informal, and institutional and official, between the three countries, but it also reflects the underlying similarities of the housing problem in London, Berlin and Paris where adequate housing simply lay beyond the means of the majority of the working classes. The artisan might be able to house his family in acceptable conditions, but the housing available to the semi-skilled and the unskilled, the vast mass of the working class, was judged by contemporaries to be expensive, insanitary and overcrowded. The essential nature of the housing problem does not change before 1914; during the period of our study the costs of decent housing remain too high for the typical

---

Source: Nicholas Bullock and James Read, *The Movement for Housing Reform in Germany and France 1840–1914*, Cambridge: Cambridge University Press, 1985, pp. 2–6, 10–11.

working-class family. In its simplest terms the challenge that faced the reformers was: how could salubrious housing be provided in a convenient location at rents that the worker could afford? As a comparative study demonstrates, the approaches adopted to solve this problem, the programme of reform, had much in common in all three countries.

The documentation of the housing reform movement in England is now well established: the studies of Gauldie, Sutcliffe, Swenarton, Tarn, Wohl and others already provide a variety of different approaches to the subject.[1] For this reason we have chosen to concentrate on a comparative study of the developments in Germany and France. In doing so we do not ignore the achievements of the reformers in England. Certainly in the early years of the housing reform movement the ideas and practice in England were better known abroad than ideas and practice in either France or Germany; frequently the debate in England precedes that in either Germany or France. But instead of offering yet another account of developments in England we have discussed the way in which English ideas and experience were incorporated from the start in the German and French discussion of the central themes of housing reform.

Our account of developments begins with the growing sense of anxiety over housing conditions. Long before the housing reform movement there had of course been housing without drainage, water supply, ventilation or light; dwellings in which whole families were crowded into single rooms were nothing new in the 1840s and 1850s; despite the protests of the reformers the situation in London in the 1840s was probably better in many respects than it had been during much of the eighteenth century. But what Edwin Chadwick and others in England had done when they first 'discovered' the 'sanitary problem' and the 'housing problem' was to force these issues before the public's attention and to demand that something be done about it. Now there was grudging recognition that housing might no longer be just a matter of concern for the individual, but did indeed touch the interests of the whole community. Poor housing undermined the family and endangered home life, and this, by extension, threatened society as a whole. The loss of moral constraint might lead to the overturning of that fragile balance between the primitive urges of man and the civilising influence of manners, order and culture; dissatisfaction with the existing order of affairs might lead to protest and, beyond, to riot and revolution; ill-health, brought on by insanitary housing, might deprive the employer of a productive workforce or the general of a fit and able-bodied fighting force.

The insanitary housing conditions in London that were already attracting public attention in the 1840s were also to be found in Paris and Berlin, but there are local variations in the 'timetable' for housing reform in each country, particularly in the early years. In Britain the publication of Chadwick's *Report on the Sanitary Condition of the Labouring Population of Great Britain* in 1842, marked a turning point in public attitudes to conditions in which so many were forced to live. More than any preceding investigation, the case presented in the report, illustrated with a mass of vivid detail, drew attention to the existence on a national scale of conditions that most people comfortably believed to be confined to the poorest areas of the largest cities. In France the publication of Villermé's *Tableau de l'Etat Physique et Moral des Ouvriers dans les Manufactures de Coton, de*

Laine et de Soie dates from 1840. But in Germany public interest in housing problems comes later. V.A. Huber's attempts to publicise the inadequacy of the housing of the working classes in the late 1840s found little response and the question did not attract widespread interest until the 1860s.

Yet, such initial differences apart, there are fundamental similarities between the development of concern for housing in all three countries: from the 1860s onwards housing is a subject of widespread public concern. Most important, in all three countries the 1880s mark another turning point in attitudes to the problem. In England, the Royal Commission on the Housing of the Working Classes reported in 1885; in Germany, the Verein für Sozialpolitik published its investigations of housing conditions in 1886; in France reformers organised the first international congress of cheap housing in 1889, and formed a national organisation, the Société Française des Habitations à Bon Marché, in the same year. All are evidence of a new attitude to social issues of which housing was a part, and also themselves gave rise to new initiatives to improve housing conditions. Earlier, solutions to the housing problem had emphasised the importance of sanitary reform and the contribution that might be made by the model dwelling companies. The conventional wisdom of the 1860s stressed the extent to which poor housing conditions were a product of the failings of the individual; as Octavia Hill put it, 'Character is the key to circumstance'. But now in the mid-1880s, in place of the almost universal belief in the efficacy of self-help, there is a recognition that the problems of housing are essentially of an economic nature, and flow from the unequal workings of society. As a result there is a new emphasis on the need for government action as the only means of changing this state of affairs. From the late 1880s onwards reformers in all three countries look to legislation as one of the key means to secure improvement.

Not only are there similarities in the timetable of the development of the movement in all three countries, but the very similarity of the nature of the housing problem in all three countries leads the debate to focus on the same basic issues. It is the major common themes of this debate, the central elements of the programme of reform and the contrasting way in which they are treated in Germany and France that form the core of our study. What, then, were the central concerns of the housing reformers?

First, there was agreement amongst all housing reformers on the vital importance of the home: it was the very foundation of a strong family life, the basis of a sound society. Well housed, the worker would be healthy, industrious, disciplined, and conscious of his stake in the existing order of society. Ownership of one's own home would further strengthen these sentiments, Samuel Smiles argued: 'the accumulation of property has the effect which it always does upon thrifty men; it makes them steady, sober and diligent. It weans them from revolutionary notions and makes them conservative.'[2]

But how was this housing to be provided? What was the ideal form of dwelling? The vision of the home for which the reformers campaigned was essentially the product of the social attitudes of the middle-class values held by the reformers themselves. In place of the reported facts of working-class life, the casual and squalid intimacy of overcrowded tenements which corroded virtues such as honesty, thrift or temperance, the reformers championed the self-contained dwellings as the key to strengthening the family. The cottage, set in its own garden, was invariably held up as the most desirable kind of home. It offered the healthiest form of dwelling, a secluded setting for the vital intimacies of family life and, at least in theory, even the possibility of home-ownership. Yet how relevant was this ideal to the vast mass of the working classes for whom it was intended? No doubt many would have chosen such a form of housing had it been accessible, but to those faced with the need to find housing within reach of work, or the necessity of taking in lodgers just to pay the rent, the reformers' ideas must have appeared an unattainable ambition.

In practice, from the very earliest days of the movement, reformers recognised that the salubrious tenement was the only practicable means of providing the ideal of the self-contained dwelling at reasonable rents in the city. Properly designed, even the tenement could provide a suitable setting for family life. Indeed the particular form of housing, whether cottage or tenement, was of less importance in creating the home than a number of other factors. Ultimately it was the self-contained dwelling, one for each family, and each with its own kitchen and hearth, that was agreed by all housing reformers as the first necessity of civilised life.

Closely related to theories about the ideal form of housing, and a second major concern of reformers, was the provision of housing that would not be injurious to health. The connection between ill-health and poor housing was recognised early on, and in England, Germany and France, anxiety over public health was one of the first and most powerful stimuli to improve housing conditions. To preserve the health of the community, sanitary and housing reformers argued that it was quite legitimate for government to intervene, a view even upheld by those like Nassau Senior generally associated with firm opposition to any extension of government activity. Indeed government might set aside the 'selfish' property rights of the few to protect the people, on the grounds argued by Shaftesbury: 'Salus populi suprema lex est'; by the end of the 1860s this principle was established in England by the Sanitary Act (1866) and the Torrens Act (1868).

Although practice might still lag behind legislative principle, the influence of sanitary reform on the housing debate can be summarised under two broad headings. First, it led to the demand for higher standards in the layout and the construction of new housing; second, it resulted in increasingly vocal demands for action to be taken to inspect and, if necessary, to regulate the use of existing housing. Calls were made for powers to enable the sanitary authorities to control overcrowding, to improve insanitary housing and, where this was not practicable, to demolish the offending properties.

In all three countries these common aims were pursued by generally similar measures: tougher building regulations and bye-laws to enforce higher standards in new housing, tighter controls of the use of existing property through a system of housing inspection, backed by tougher sanitary regulations. Naturally there are differences of emphasis, a product of the different patterns of urban growth, and the different traditions of housing from one country to another: in neither France nor Germany was there any equivalent to national legislation such as the Cross Act (1875) which gave the local authorities powers to order the large-scale demolition of insanitary housing. Yet, even without the powers conferred by national legislation, early successes could be matched by later developments: in all three countries sanitary reformers could legitimately point to a number of lasting achievements. In Germany, in particular, beset with the problems of very high density tenement housing, the sanitary reformers campaigned long and hard for tighter control over the proportion of the site that might be developed and the maximum bulk of a building. The fruits of this campaign, the zoning ordinances successfully applied in many German cities in the mid-1890s, were to have important consequences for later developments in planning both in Europe and in America. [. . .]

The housing reformers may have done too little too late, their achievement in practice may be belittled as no more than the provision of better housing for a relatively prosperous few, as part of a programme which they hoped would gradually filter down to the many. But this is to overlook one of the movement's most significant achievements. Whatever the limitations in practice, their ideas, investigations and discussions were of lasting importance. Housing reformers investigated the economics of housing, they explored the design and the hygiene of the dwelling; they demanded a consideration of the problems of housing as part of the growth of cities and the expansion of the urban population. These investigations and the understanding of the problems of housing that they established were of immense importance after 1914. In England, in France and in Germany, the form of government intervention, and thus the subsequent history of housing and housing policy, was crucially determined by the debates and attitudes of the pre-war years. If we are really to understand the making of housing policy since 1914, we must look at the activities of the pre-war movement for housing reform.

Notes

1 The recent literature on housing reform in England is extensive, see for example: E. Gauldie, *Cruel Habitations, a History of Working-Class Housing 1780–1918* (London, 1974); A. Sutcliffe (ed.), *Multi-Storey Living: The British Working-Class Experience* (London, 1974); M. Swenarton, *Homes Fit for Heroes, the Politics and Architecture of Early State Housing in Britain* (London, 1981); J.N. Tarn, *Five Per Cent Philanthropy, an Account of Housing in Urban Areas between 1840 and 1914* (Cambridge, 1973); A.S. Wohl, *The Eternal Slum, Housing and Social Policy in Victorian London* (London, 1977).

2 S. Smiles, *Thrift* (London, 1875).

4.6 Donald MacKenzie, *Eugenics in Britain*, 1976

The eugenics[1] movement, which flourished in Britain around the early part of this century is an important example of the relationship between scientific ideas and the interests and purposes of social groups. The eugenists possessed a social theory, and a set of social policies, which claimed scientific foundation. Social position, they argued, was largely the result of individual qualities such as mental ability, predisposition

Source: *Social Studies of Science*, vol. 6, 1976, pp. 499–532.

to sickness or health, or moral tendency. These qualities were inherited, and thus a rough equation could be drawn between social standing and hereditary worth. On this basis a programme of social action to improve the quality of the population was put forward. Central to this was the alteration of the relative birth-rate (or survival rate) of the 'fit' and 'unfit'. Those with good hereditary qualities should marry with care and have large numbers of children (this came to be called positive eugenics), while those with hereditary disabilities should be discouraged from parenthood (negative eugenics). The eugenists supported schemes of social reform which would, either directly or indirectly, have this effect, while condemning policies which appeared to encourage procreation of the 'unfit'. Thus, they sought to raise the fertility of some groups in society (generally those of higher social status) and lower that of others (those of lowest status).

Eugenics was backed by arguments based on commonsense and medical knowledge of heredity, Darwinian biology and, increasingly, specialized scientific research. While largely relying on pre-existing ideas of society and human heredity, the eugenists themselves developed a body of knowledge of direct eugenic relevance.[2] Much of this has become integrated into modern science. Eugenists played the crucial role in the development of mathematical statistics in Britain, through the work of Francis Galton, Karl Pearson and their collaborators, and their ideas informed much early work in genetics. Psychological testing and psychometric theories were developed primarily by men with eugenic convictions (Galton, Charles Spearman, Cyril Burt). Thus, the science of the eugenists made a considerable impact on the scientific and intellectual development of twentieth century Britain.[3] Further, the content of this science (the concepts, theories and methods used) was in large part determined by the eugenic purposes for which it was developed. [. . .]

A few initial words on the perspective of this paper are perhaps desirable. I shall attempt to explain the rise and decline of eugenics, some aspects of the content of eugenic ideology, and the differential appeal of eugenics to the various social classes and occupational groups in Britain. British society I see as fundamentally divided between capitalist and residual aristocratic groups on the one hand ('the ruling class'), and the manual working class on the other. However, by the late nineteenth century important intermediary groups had appeared, notably the new professional occupations (school-teaching, science and engineering, etc.). Together with the established professions of the church, the law and medicine, these formed what is conventionally and usefully referred to as 'the professional middle class'. Within this group there were, however, important divisions – for example in the nature of the specialized knowledge which legitimated the particular professional role. The church and the legal profession relied on traditional spheres of knowledge, while many of the newer professions (and increasingly the older profession of medicine) sought legitimations in such fields as natural science and empirical social research.[4]

I shall argue that eugenics should be seen as an ideology of the professional middle class, and in particular of the 'modern' rather than the 'traditional' sector. Eugenic ideas were put forward as a legitimation of the social position of the professional middle class, and as an argument for its enhancement. At the same time the eugenic programme was seen by its protagonists as a solution to the most pressing perceived problems of social control in British society. It was thus put forward as a strategy for the ruling class, and the plausibility of eugenics as such a strategy is an important variable in explaining its rise and decline. Eugenic ideas can be regarded as a set of tools deployed for social purposes. The ideas were taken up when thought likely to be useful to their carrier group, and later, when changed circumstances made them less appropriate, they were discarded.

No attempt will be made to compare the eugenic movements of different societies. The analysis offered is particular to the British situation; only its general assumptions and perspectives could be applied in other contexts. British eugenics was unique in many respects. In particular, it was a class rather than a 'racist' phenomenon, and unlike its German and United States equivalents is not to be understood in terms of preoccupation with Jews, Blacks or immigrants. Doubtless British eugenists, like Britons in general at this time, held 'racist' views, but these prove largely incidental to their eugenic concerns.

The background to eugenics

The eugenists did not develop their ideas in an intellectual vacuum. They were able to draw on pre-existing beliefs about heredity and society. They fashioned their theory in accord with their purposes by taking some of these beliefs, transforming some of them, and adding new elements.

In his biography, Galton described early nineteenth-century beliefs about heredity as 'lax and contradictory'.[5] To the extent that this was so it can be attributed to the large variety of social purposes which such beliefs served. The animal breeder used heredity as a guide in developing stock; the physician used it as an explanation of disease; the moralist used it to sanction deviance; the middle class male used it as an argument for female passivity. Before eugenics there was no single dominant social use to which heredity was put, there was no generalized controversy about heredity, and thus there was little pressure to consistency in the deployment of ideas. 'Clarification' came only as a result of the eugenists' systematic and controversial use of the ideas of heredity; pre-eugenic notions formed a rich, varied and plastic body of knowledge capable of easy deployment in various directions.[6]

Hereditarian beliefs *were* invoked in arguments about social reform before eugenics, but the use made of them was frequently opposite to that typical of the eugenics movement. Heredity could be invoked as a sanction reinforcing the case for particular environmental reforms. Thus, bad conditions, drunkenness and drug abuse were held to have a detrimental effect on the children of the present generation through the inheritance of acquired characteristics. Environmental reform – sanitary improvements, a curb on the drink trade – would arguably improve not simply this generation but the next.[7] [. . .]

It is not possible to attribute the change in the social uses of hereditarian beliefs in the later nineteenth century simply to internal changes within science. Certainly, most British biologists after 1890 did follow August Weismann in his rejection of the view that acquired characters could be inherited. And eugenists did use this as a basis for arguing that only eugenic reform could have a permanent effect on the race. However it is clear that Weismannism did not *cause* eugenics. Galton had independently rejected the inheritance of acquired characters before Weismann's work appeared, probably because of his eugenic views.[8] And the subsequent reception of Weismann's views in Britain was strongly conditioned by their perceived political significance.[9] There had, in fact, been no major change in the available scientific evidence. Nor did acceptance of Weismannism compel or even indicate advocacy of eugenics.[10]

Another component in the intellectual background of eugenic thought was political economy and the image of society it developed.[11] For all its rejection of Enlightenment optimism, of the environmentalism of the utilitarians, and of the revolutionary-bourgeois notion that 'all men are born equal', eugenics retained certain key elements of classical bourgeois thought. The eugenic view of society was individualistic and atomistic. The fitness of a society was the sum of the fitnesses of the various individuals comprising it. Although the eugenists stressed race, their view of race was not a holistic one. The race was not an unalterable essence, but an historical population, the sum of its parts.

There was a particularly close affinity to the biological variant of political economy, social Darwinism. The eugenic identification of social failure with biological unfitness, the notion of progress coming through the elimination of the unfit, and the biological view of society, are all drawn from social Darwinism. Earlier social Darwinism (especially Spencer's) held that the elimination of the unfit could be achieved by political inaction. If the state would stop interfering in the working of natural laws, all would be well.[12] Eugenics, in contrast, did not trust to laissez-faire. 'What Nature does blindly, slowly, and ruthlessly', wrote Galton, 'man may do providently, quickly, and kindly'.[13]

Thus, eugenic thought drew on resources present in the culture of Victorian Britain. But it combined these in its own characteristic manner and, in addition, developed patterns of thought of an entirely novel kind: both general, such as the nature/nurture distinction, and more specialized, such as the statistical view of heredity and evolution. We must now consider who developed and propagated this new and characteristic body of thought.

The social composition of the eugenics movements

British eugenics can, for our purposes, be said to have begun in the 1860s with the publication of the first articles on the subject by Galton and Greg. During the 1880s eugenics became a definite topic of public discussion in books and articles. Between 1900 and 1914 it achieved institutional expression, notably with the establishment of a Eugenics Laboratory in the University of London and in 1907 with the foundation of the Eugenics Education Society (EES). By 1913–14 the EES had over 1,000 members.[14]

The most straightforward answer to the question, 'Who were the eugenists?', is provided by examining the membership of the EES in the key years 1908–14. With the exceptions (notably Karl Pearson), nearly all known active British eugenists seem to have been members of the Society. Its membership and activities have been documented by Farrall. His investigations lead him to conclude that:

> The leadership of the Eugenics Education Society was dominated by well educated members of the middle-class professions of medicine, university teaching and science . . . Membership was not only drawn almost exclusively from the middle classes but also heavily from the intellectual, creative and welfare professions. Of those whose profession has been discovered only three military officers and one businessman would be excluded definitely from this category.[15]

[. . .] [T]he general tenor of the evidence supports the association of eugenic thinking with the professional middle class. Let us see whether the content of eugenic thought can be said to reflect the social base of the eugenics movement. Can we impute eugenics as an ideology of the professional middle class? If so, the empirical association of class and eugenic attitude becomes of greater sociological significance.

Eugenics as an ideology of the professional middle class

Let us begin by examining Francis Galton as the founder of British eugenics, before turning to the movement as a whole. What was the relation between Galton's eugenics and his social experience?

By birth, marriage and inclination Galton belonged to the élite of the Victorian professional middle class. N.G. Annan has called the group to which Galton belonged 'the intellectual aristocracy'.[16] The origins of this group lay in the bourgeoisie. The families from which this group came were distinguished from the bulk of the bourgeoisie by religion (they were Quakers, Unitarians or members of the Clapham Sect) and by their philanthropic and anti-slavery concerns. The children of marriages within this group tended to abandon direct business involvement for the world of scholarship, education and the professions. They rapidly rose to dominant positions in the universities, public schools, science and literature. Some entered the state bureaucracy, to become 'mandarins' of the increasingly professional civil service. Although links of kinship and common interest bound them to other sections of the élite of Victorian Britain, Annan emphasizes that this group maintained a separate identity.[17] At least until the end of the nineteenth century it remained tightly-knit, held together by continuing intermarriage and by a common commitment to educational modernization and administrative reform, to the abolition of religious tests and to the introduction of selection by competitive examination in the civil service. 'The intellectual aristocracy' stood for change in British society, but for change that was gradual and piecemeal, that would be achieved by argument and persuasion within the 'corridors of power'.[18]

Francis Galton could well be taken as an archetype of this group. It seems that direct observation of kinship links within this professional élite was the source of his initial hereditarian convictions. He did not, however, give any general interpretation to this to begin with. The spur to such an interpretation was the publication by his cousin. Charles Darwin, of *The Origin of Species*. Fifty years later, Galton wrote:

> The publication in 1859 of the *Origin of Species* by Charles Darwin made a marked epoch in my own mental development, as it did in that of human thought generally. Its effect was to demolish a

> multitude of dogmatic barriers by a single stroke, and to arouse a spirit of rebellion against all ancient authorities whose positive and unauthenticated statements were contradicted by modern science.[19]

As important as any detailed impact Darwin's work had on Galton was the more general effect on him of the controversy following its publication. Galton was present at the British Association meeting at Oxford in 1860 when Huxley and Wilberforce debated Darwin's theories.[20] Galton clearly felt the need to choose sides between scientific naturalism and its theological opponents:[21] given his background, there could be little doubt which side he would choose. He became a leading member of the group of scientific intellectuals which included Huxley, Spencer and Tyndall. He vigorously opposed the dogmas of revealed religion, and sought to replace the Christian faith by a system of belief based on natural science.[22] The near monopoly of the church in comfortable professional positions must be ended, and an adequately-supported profession of science established. The scientists' role should not be a mere technical one: they should form

> a sort of scientific priesthood through the kingdom, whose high duties would have reference to the health and well-being of the nation in its broadest sense.[23]

In the 1860s Galton began to interpret his experience of kinship links in the professional élite in a naturalistic and evolutionary framework, and to derive from this a faith and a social practice for the scientific priesthood. He sought to trace kinship links amongst those acknowledged to be of exceptional mental ability (amongst his examples were the Darwin and Butler families). By this means he showed that achievement ran in families (i.e. the closeness of kinship links amongst the eminent was far greater than would be expected if eminence was distributed at random in the population). This he interpreted as proof of the inheritance of mental ability, and on this basis he argued for a eugenic programme which would insure the careful and early marriage and high fertility of the most able.[24] For Galton, eugenics was not a mere minor reform. He saw in eugenics the basis for a new scientific and evolutionary religion, in which an individual would be seen only as a manifestation of immortal germ plasm. This new faith implied the dominance of the 'scientific priesthood' over revealed religion. The practice of eugenics also necessitated social changes. The dominance of society by plutocracy and hereditary nobility must be ended. Extremes of inherited wealth and titles of nobility had a bad effect on the race, causing the degeneration and sterility of originally healthy stock. Instead,

> The best form of civilisation in respect to the improvement of the race, would be one in which society was not costly; where incomes were chiefly derived from professional sources, and not much through inheritance; where every lad had a chance of showing his abilities and, if highly gifted, was enabled to achieve a first-class education and entrance into professional life, by the liberal help of the exhibitions and scholarships which he had gained in his early youth.[25]

Galton's eugenics had thus a double aspect. It was expressive of his social experience. He came from an intellectual élite closely bound by kinship ties. In his social group achievement *was* inherited (though we might now want to interpret this socially rather than biologically). Successful fathers had successful sons, these sons generally married within the social group and themselves had successful offspring.[26] So Galton was interpreting generally and naturalistically a salient facet of his social experience. At the same time, an intrinsic part of the eugenic programme was the advancement of the interests of the professional middle class. The middle class 'expert', rather than the priest, aristocrat or plutocrat, should exercise power in an efficient modernized eugenic society. Science, rather than Christian religion, should be the dominant cultural form. Thus, Galton's eugenics, as well as expressing his social experience, also represented the interests of his social group.

Although Galton was the founder of the eugenics movement, the analysis of his work alone does not establish the nature of eugenics as a whole. I do not wish to argue that eugenics remained expressive of the social experience of the eugenists in the sense that this was true for Galton. Later eugenists were generally of a lower status within the professional middle class than the 'intellectual aristocracy', and few would have had such strong kinship links to the élite as Galton had. On the other hand it remains true that eugenic thought expressed the interests of the professional middle class, both in a narrow sense and in the wider sense of the relative status of professionals and other middle and upper class groups.

At times the Eugenics Education Society acted as a straightforward advocate of the financial interest of the middle class:

> . . . the incidence of the income tax is claiming attention, and a letter has been sent by the President to all Members of Parliament pointing out that any system of taxation which takes no account of the necessary expenditure involved in bringing up a family may, in a sense, be said to penalize marriage and parenthood, and that taxation which retards marriage and discourages parenthood on the part of worthy citizens has a harmful influence in tending to lower the proportion of men and women of good stock or blood in the composition of the generation of the future. There is no question that the income tax at present falls most heavily on parents belonging to the middle and professional classes, to whom the description can be approximately applied.
>
> It is suggested that the way to remedy this evil is to extend the principle of allowing rebates for each child . . .[27]

When the First World War broke out the Council of the EES discussed what practical eugenic action could be taken in the war situation. As a result of this discussion the EES, in conjunction with the heads of the leading professional bodies and institutions, helped form a Professional Classes War Relief Council and set up a maternity home for the wives of professional men serving in the armed forces.

The significance of eugenics as regards the professional middle class was, however, much wider than this. Eugenics served both to legitimate the social position of the professional middle class and as an argument for its improvement. The professional middle class owed its social position neither to wealth nor to ascribed status but to the specialized mental abilities and knowledge of its members. The hereditarian theory of mental ability as developed by the eugenists claimed that only a limited section of the population had the potential to achieve the skills and knowledge required for professional middle class roles. The professional middle class had achieved their position not by accident of circumstances, but as the result of generations of selection for mental ability. The next generation of professionals would of necessity have to be recruited from the middle class. Thus a rigidly stratified educational system was justified, with only the narrowest of ladders to allow the unusually gifted child, the 'sport', to rise from the lower classes. Eugenics offered the professional middle class an educational philosophy which enabled them to justify the effective monopoly of professional education by the existing professional class. The eugenist could consistently advocate an expanded educational system – 1870–1914 was a period of considerable educational expansion – while laying down a structure for this expansion which maintained existing privileges.

One interesting facet of the discussion of mental ability by British eugenists is that 'business acumen' or 'entrepreneurial skills' played no part in it. We find no 'English Men of Business' paralleling Galton's *English Men of Science*, although an hereditarian account of business skills could have been constructed with equal plausibility. While the majority of British eugenists did not attract the business community, they did not seek to legitimate it in a similar way to their legitimation of the professional middle class. There was also no attempt to legitimate the hereditary nobility. Indeed a not uncommon target for attacks by eugenists was the House of Lords. Following Galton's views on the detrimental effect on the race of the peerage, schemes such as the replacement of the House of Lords by an Upper House of families of genuine eugenic worth were discussed. Arnold White, for example, pictured the aristocracy and plutocracy as degenerate and prey to hereditary ills as the result of inbreeding and marriage for wealth rather than for health and mental ability.[28]

The majority of eugenists stopped short of an explicit attack on the existing power structure of British society. A significant section, however, attacked the existing ruling groups as unable to administer a modern society efficiently and scientifically and condemned capitalist society as dysgenic (i.e. anti-eugenic) in its operation. A eugenic policy, they argued, was impossible while laissez-faire capitalism demanded large supplies of unskilled labour and a permanent pool of unemployed. Among these 'socialist' eugenists were Karl Pearson, Jane Hume Clapperton and several leaders of the Fabian Society, including Sidney Webb, George Bernard Shaw and (for a while) H.G. Wells. [. . .]

Eugenics, the residuum and social imperialism

Eugenics was not only a matter of raising the fertility (and status) of the professional middle class; it also involved lowering the fertility of those at the bottom of the social scale. While this aspect of it was little emphasized in Galton's early, utopian, positive eugenics, it came more and more to the fore in the period from 1880 onwards. Within Galton's own work negative eugenics became more prominent (though he always treated the subject with a certain caution, even distaste, and avoided 'unmentionable'

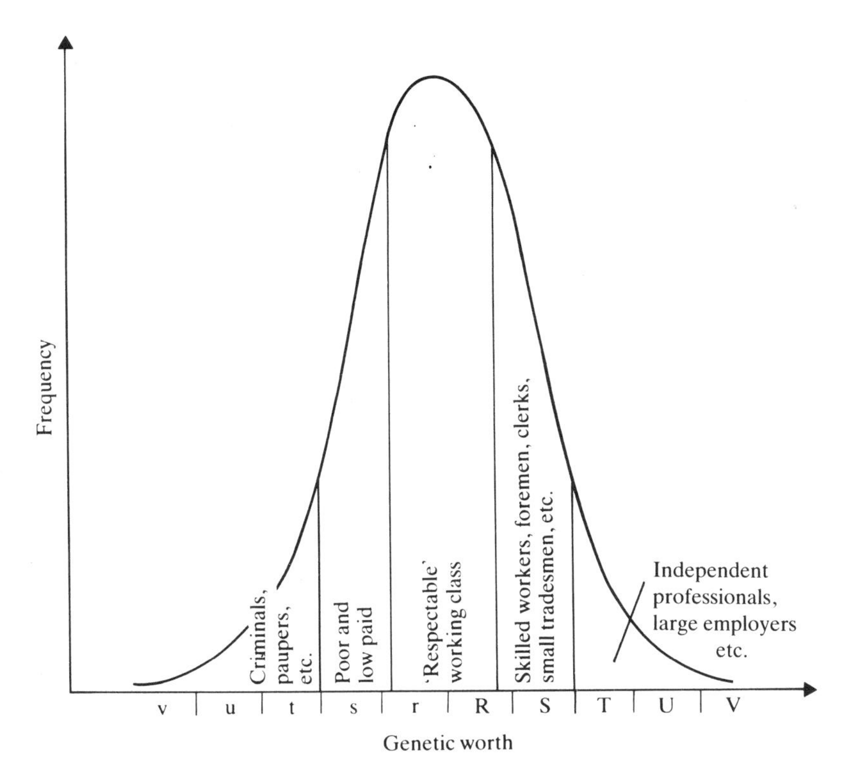

topics such as sterilization and contraception). More generally, the 'unfit' rather than the 'fit' were the central focus of eugenic propaganda. What, we must ask, were the views on class structure held by the eugenists, and who were the unfit who were to be dissuaded from breeding?

The eugenists accepted a rough equation of social standing and genetic worth. Indeed, this is generally an axiom with their thought, seldom a proposition they feel any need to defend. At least for those social groups conventionally regarded as being below the professional middle class, class position was taken as a sure indicator of average mental ability. The view of social structure they held was summarized by Galton in his 1901 Huxley Lecture.[29] Galton took the social categories of Booth's survey of London and mapped them on to his assumed distribution of inherited 'civic worth'. In Figure 1, I have presented his results in graphical form. 'R, S, T, U, V' and 'r, s, t, u, v' are the subdivisions of 'civic worth'. The lowest group, classes t, u, v and below, 'are undesirables'.[30] It is against them (and particularly against the 'criminals, semi-criminals and loafers' of v and below) that negative eugenics should be practised; for example, habitual criminals should be segregated 'under merciful surveillance' and 'peremptorily denied opportunities for producing offspring'.[31] Galton (and the other eugenists) did not wish to depress the birth rate of *all* groups below the middle class. It would scarcely have been in the interests of the middle class to do so! All eugenists were agreed that manual workers were socially necessary. What they wanted was to improve the discipline, physique and intelligence of the working class by eradicating the 'lowest' elements of it. The eugenists attempted to draw a line between socially useful and socially dangerous elements of the lower orders. While the exact placing of this line was vague, and varied from one writer to another, all were agreed that this distinction was necessary. In few cases was the view of social class as explicit as it was in Galton's writings: nonetheless, all eugenists would have adhered to a similar model. Indeed, few members of the middle class of Victorian and Edwardian Britain would have found much to disagree with in Galton's model. The specificity of eugenic thought lay not in the model, but in the conclusions for action drawn.

The lowest social group ('t, u, v and below') were a prominent – indeed *the* prominent – social problem in the eyes of middle class late Victorians and Edwardians. The attitudes of the middle class to this group have recently been elucidated by Gareth Stedman Jones in his *Outcast London*.[32] Jones argues that in the latter part of the nineteenth century the focus of middle class fears about social stability, doubt about industrialism and urban existence, shifted from the heartlands of the industrial revolution (such as Manchester) and became centred on London. Since the decline of Chartism, most middle class observers felt that the respectable working class of the North of England were no longer a threat or a social problem. The problem rather lay with a smaller and more specific group on the slums of the big cities.

> The most characteristic image of the working class was that of increasingly prosperous and cohesive communities bound together by the chapel, the friendly society, and the co-op. Pitted against the dominant climate of moral and material improvement was a minority of the still unregenerate poor: those who had turned their backs on progress, or had been rejected by it. This group was variously referred to as 'the dangerous class', the casual poor or most characteristically, as the 'residuum'.[33]

In other words, the perceived problem of social control was no longer the working class as a whole, but only a residual section of it. [. . .]

The residuum posed a problem of social control. They were not, it is true, radical or revolutionary. But they were politically volatile, and, pressed by extreme hardship, they were liable to riot.[34]

The middle class were not concerned with social control alone. They felt that the poor were not only dangerous but also physically and mentally degenerate. Characteristically, the urban slum dweller was compared with the healthy and strong agricultural labourer. It was widely believed that urban conditions caused the degeneration of immigrants from the country, whether by the direct effect of environment or by selection of the worst types. Francis Galton was an early proponent of the theory of urban degeneracy:

> It is perfectly distressing to me to witness the draggled, drudged, mean look of the mass of individuals, especially of the women, that one meets in the streets of London and other purely English towns. The conditions of their life seem too hard for their constitutions, and to be crushing them into degeneracy.[35]

Increasingly, the context in which the problem of urban degeneration was seen, was that of imperialism. A degenerating population was serious enough under any circumstances, but it would be fatal to a British Empire faced with increased foreign economic competition, colonial war and the ultimate threat of inter-imperialist war. The early reverses suffered by British troops in the Boer War (1899–1902) gave concrete form to these misgivings. It was put about, and widely believed, that up to 60 per cent of working class volunteers for the army had had to be rejected because they failed to meet the army's minimum standards of physical fitness.[36]

The problem, then, was seen to be a selection of the working class that lacked both moral fibre (i.e. was outside social control) and physical fitness. The urban situation had broken the older forms of social control based on direct personal contact between rich and poor. The most important early attempt at a solution was the Charity Organisation Society, set up in 1869, which sought to reimpose social control through organized, selective charity and trained social workers. But with the deepening urban crisis of the 1880s and the serious rioting, there was a conscious search for new responses to the problem. Crucial to these responses was the distinction between the respectable working class and the residuum: the residuum must be isolated from the working class as a whole (even at the price of concessions to the bulk of workers) and neutralized or eliminated. The growing awareness of competition between imperialist powers underlined the urgency of the problem. A modern imperialist state needed an efficient, fit and loyal working class. As the riots of the 1880s and the debacles of the Boer War indicated, there was a weakness at the very heart of the British Empire. Fabians, Tories and Liberal Imperialists could find common ground in agreement that a solution to the problem of the urban residuum was a prerequisite of imperial survival. The basis was thus laid for social imperialism, the linking of imperialism and social reform that loomed large in British politics between the 1880s and 1914, and which, as Farrall points out, was important to eugenic thought.[37]

It was in this context that eugenics provided a plausible social policy. The eugenists had a biological explanation of the residuum. The suspension of natural selection through the operation of charity, medical science and sanitary reform had led to the flourishing in the hearts of the great cities of a group of people tainted by hereditary defect. They were unemployed because they lacked the health, ability and strength of will to work. Hereditary weakness turned them to crime and alcohol. Their constitutions inclined them to wasting diseases such as tuberculosis. This group of degenerates was outbreeding skilled workers and the professional middle class. Further, the eugenists warned, although natural selection was largely suspended *within* British society, competition between different nations went on. Britain was engaged in a struggle for survival that was at present commercial but might become military. National fitness for this struggle was necessary. This had previously been ensured by natural selection, but under the conditions of modern civilization a replacement for natural selection had to be found in conscious eugenic selection. A pliable and fit working class could be bred by isolating the residuum in institutions where parenthood would be made impossible.

Negative eugenics was thus not an abstract programme, but a specific response to a specific problem. The eugenists proposed the most thorough solution to the problem of the residuum short of immediate elimination. Social control was to be imposed by the detention in institutions of the habitual criminal, the alcoholic, the 'hereditary' pauper, and so on. Prevention of parenthood in these institutions would mean the eventual disappearance of the residuum as a group. This solution would leave untouched the position and privileges of the higher social classes, while drawing in full on the skills of the middle class scientific expert. While it might seem a rather extreme proposal, it differed only in thoroughness and scientific rationale from similar proposals put forward at the time, for example, for labour camps with compulsory powers of detention (proposals that were supported by Fabians and 'humanitarian' Liberals).[38]

The rise and decline of eugenics

The rise and decline of the eugenics movement in Britain seems to be largely accounted for by variations in the credibility of the programme for negative eugenics. Four major turning points can be identified: the sense of an urban crisis in the 1880s, the Boer War (1889–1902), the First World War and the world slump and the emergence of German fascism (1929–34).

Before 1880 it is impossible to talk of eugenics as a movement: it must have seemed to be a utopian speculation. The urban crisis of the 1880s and the related emergence of social imperialism and Fabianism provided the context for serious consideration of negative eugenics.[39] The real opportunity for the eugenists came with the Boer War and the boost it gave to social imperialism. This prompted Karl Pearson and Arnold White to write their most famous social imperialist and eugenic tracts.[40]

Pearson wrote to Galton urging him to open a direct campaign for eugenics, sensing that the time was ripe for 'a word in season' on eugenics. Although almost in his eighties, Galton responded by campaigning for, funding and supporting eugenics. The years from 1901–1914 were of almost uninterrupted success for the eugenics movement, which by the time of the outbreak of war seemed on the threshold of considerable legislative impact. Prominent political figures had at least shown interest in eugenics, as was witnessed by the presence of names such as A.J. Balfour and Winston Churchill in the list of Vice-Presidents of the International Eugenics Congress held in London in 1912. A small but growing group of MPs responded to eugenic ideas, and the Eugenics Education Society was able to claim the formulation and passing of the Mental Deficiency Act of 1913 as the result of its work.[41]

After 1918 all this impetus had gone. There was no disastrous immediate decline of British eugenics. The cadre of the movement remained intact. But eugenics seemed to lack political credibility. The EES (renamed the Eugenics Society) evolved gradually into a learned society rather than a campaigning political group. The broad spectrum of political support in the professional middle class evaporated. Increasingly, eugenics as a full-scale political programme became identified with the extreme right-wing. What went wrong for the eugenists?

The answer appears to be that the conditions for the credibility of the social programme of negative eugenics no longer existed after 1918. Before the War the problem of social control was seen as centred on a relatively small and well-defined subgroup of the working class. After 1918 things were different. Red Clydeside and the industrial battles of the 1920s suggested that there was a pressing danger to established society from the working class as a whole. Unemployment was no longer localized (indeed London, the core of unemployment before 1914, was relatively prosperous during the 1920s and 1930s by comparison with the industrial North). A political strategy for the British ruling class clearly had to involve a reckoning with the working class as a whole. Such a strategy did evolve, empirically rather than theoretically, in the 1920s. Although it involved intransigence at certain key moments (notably the General Strike of 1926), the key to the strategy was an accommodation with the political and industrial leadership of the working class in the Labour Party and trade unions. This left no place for eugenics; for example, to make the point starkly, sterilization of the unemployed (as advocated by E.W. MacBride[42]) was out of place in such a strategy. It was impossible both to reach a compromise with the official leadership of the working class *and* to threaten that class (or a significant subsection of it) with negative eugenics.

Most eugenists gradually came to terms with this reality and diluted their proposals accordingly (for example by calling for voluntary rather than compulsory sterilization). Some, like R.A. Fisher, ceased to propagandize for eugenics, while continuing privately to hold eugenic beliefs.[43] A few maintained the old attitudes intact and looked to the application of eugenic measures in the context of the destruction of the labour movement rather than accommodation with it: thus George Pitt-Rivers, formerly Secretary of the International Federation of Eugenic Societies, joined the British Union of Fascists and was interned during the Second World War.[44] The Nazi victory in Germany and the subsequent Nazi eugenic measures strengthened the association of eugenics and the extreme right. After some initial hesitation, the Eugenics Society condemned Nazi eugenics.[45] But an already enfeebled Society found it difficult to make it clear that what it preached was different from what the Nazi practised. By the late 1930s eugenics in the old, strong, sense was identified with fascism. In the absence of gains for fascism within British society, eugenics was bound to decline. [. . .]

Notes

1 Francis Galton first used the term eugenics in his *Inquiries into Human Faculty* (London: Macmillan, 1883), 25. The concept was implicit from the beginning of his work in heredity 20 years earlier. The Greek root means 'of good stock'.

2 Throughout this paper I use the word 'knowledge' in the sociologists' sense of 'accepted belief' and do not wish to imply any judgement as to the truth of the ideas in question.

3 The impact of eugenics on statistics is discussed in Ruth Schartz Cowan, 'Francis Galton's Statistical Idea: the Influence of Eugenics', *Isis*, Vol. 63 (1972), 509–28; also in D. MacKenzie, 'Social Factors in the Emergence of

Modern Statistics', paper read to the Conference on the History of Statistics, Harvard University (January 1974). Cowan discusses Galton's influence on genetics in 'Francis Galton's Contribution to Genetics', *Journal of the History of Biology*, Vol. 5 (1972), 380–412. Philip Abrams, *The Origins of British Sociology, 1834–1914* (Chicago: University of Chicago Press, 1968) indicates the general intellectual impact of eugenics.

4 For an account of the growth of professional occupations in Britain, see A.M. Carr-Saunders and P.A. Wilson, *The Professions* (Oxford: Clarendon Press, 1933). In identifying particular occupations as 'professional' or not, I would follow their judgement.

5 F. Galton, *Memories of my Life* (London: Methuen, 1908), 288.

6 Cf. C.E. Rosenberg, 'The Bitter Fruit: Heredity, Disease and Social Thought in Nineteenth-Century America', *Perspectives in American History*, Vol. 8 (1974), 189–235.

7 It should, however, be noted that many of those campaigning for compulsory custodial treatment of alcoholics did not accept this optimistic view, and saw retreats as a means of isolating alcoholics. See R.M. MacLeod, 'The Edge of Hope: Social Policy and Chronic Alcoholism, 1870–1900', *Journal of the History of Medicine and Allied Sciences*, Vol. 22 (1967), 215–45.

8 Ruth Schwartz Cowan, 'Sir Francis Galton and the Continuity of the Germ-Plasm: a Biological Idea with Political Roots', *Actes du XIIe Congres International de l'Histoire des Sciences* (Paris, 1968), Vol. 8, 181–86.

9 See, for example, W.P. Ball, *Are the Effects of Use and Disuse Inherited?* (London: Macmillan, 1890). J.C. Burnham argues that this socio-political response to Weismannism was a peculiarly Anglo-American phenomenon. See his 'Instinct Theory and the German Reaction to Weismannism', *Journal of the History of Biology*, Vol. 5 (1972), 321–26.

10 A.R. Wallace, for example, accepted that acquired characters were not but rejected eugenics as a social programme.

11 Abrams makes much of the resemblance of eugenic thought to political economy: 'Eugenics, when its history is written, will have to be treated in close relation to political economy', op. cit., 123.

12 See, for example, Herbert Spencer, *The Study of Sociology* (London: Henry King, 1873), 343–46.

13 F. Galton, 'Eugenics: its Definition, Scope and Aims', in his *Essays in Eugenics* (London: Eugenics Education Society, 1909), 42. For an account of social Darwinism's broadly similar development in America, see R. Hofstadter, *Social Darwinism in American Thought* (Boston, Mass.: Beacon Press, 1968).

14 L.A. Farrall, *The Origin and Growth of the English Eugenics Movement, 1865–1925* (PhD Thesis, Indiana University, Bloomington, 1970), 211–12.

15 Ibid., 225 and 228.

16 N.G. Annan, 'The Intellectual Aristocracy', in J.H. Plumb (ed.), *Studies in Social History: a Tribute to G.M. Trevelyan* (London: Longmans Green, 1955).

17 Ibid., 248.

18 Ibid., 247.

19 Galton, op. cit. note 5, 287.

20 D.W. Forrest, *Francis Galton: the Life and Work of a Victorian Genius* (London: Paul Elek, 1974), 84.

21 See F.M. Turner's *Between Science and Religion* (New Haven, Conn.: Yale University Press, 1974) for a sensitive analysis of this divide.

22 Some of Galton's anti-clerical sallies – in particular his statistical 'disproof' of the efficacy of prayer – now seem rather amusing. But at the time issues such as this were seriously debated. See F.M. Turner, 'Rainfall, Plagues and the Prince of Wales: a Chapter in the Conflict of Religion and Science', *Journal of British Studies*, Vol. 13 (1974), 46–65.

23 F. Galton, *English Men of Science: their Nature and Nurture* (London: Macmillan, 1874; facsimile reprint, with an introduction by R. Schwartz Cowan, London: Frank Cass, 1970), 260.
24 See F. Galton, *Hereditary Genius* (London: Macmillan reprint of second edition (Gloucester, Mass.: Peter Smith, 1972).
25 Ibid., 415.
26 Mothers and daughters, it is worth noting, scarcely figured in Galton's eugenic thought except as the *transmitters* of hereditary ability.
27 EES, *Sixth Annual Report* (1913–14), 7.
28 See his *Efficiency and Empire* (London: Methuen, 1901).
29 F. Galton, 'The Possible Improvement of the Human Breed, under the existing Conditions of Law and Sentiment', reprinted in his *Essays in Eugenics*, op. cit. note 13, 1–34.
30 Ibid., 11.
31 Ibid., 20.
32 Gareth Stedman Jones, *Outcast London* (Oxford: Clarendon Press, 1971).
33 Ibid., 10–11.
34 For the most serious disturbances (those of the mid-1880s) see ibid., 290–96.
35 Galton, *Hereditary Genius*, op. cit, note 24, 395–96.
36 See Bentley Gilbert, *The Evolution of National Insurance in Great Britain* (London: Michael Joseph, 1966), 85 ff., for different responses to the scare.
37 See Farrall, op. cit. note 14; B. Semmel, *Imperialism and Social Reform: English Social Imperial Thought, 1895–1914* (London: George Allen and Unwin, 1960); and G. Searle, *The Quest for National Efficiency* (Oxford: Blackwell, 1971).
38 See Jones, op. cit. note 32, passim., for some of these proposals.
39 See, for example, Arnold White, *Problems of a Great City* (London: Remington, 4th edition, 1895), first published in 1886.
40 Karl Pearson, *National Life from the Standpoint of Science* (London: Adam and Charles Black, 1901); White, op. cit. note 28.
41 EES, *Sixth Annual Report* (1913–14), 5–6. See Farrall, op. cit. note 14, 238-47, for the political activities of the EES.
42 In *Nature*, Vol. 137 (1936), 45. Quoted by P.G. Werskey in his article 'Nature and Politics between the Wars', *Nature*, Vol. 224 (1969), 462–72.
43 Interviews with students of Fisher have convinced me of this point.
44 P.G. Werskey, 'British Scientists and "Outsider" Politics 1931–1945', in S.B. Barnes (ed.), *Sociology of Science* (Harmondsworth, Middx.: Penguin, 1972), 252. The British Union of Fascists employed eugenic rhetoric. See Sheila Rowbotham, *Hidden from History* (London: Pluto, 1973), 151.
45 See the *Eugenics Review*, Vol. 25 (1933–4) and Vol. 26 (1934–5), passim.

5
Science, Technology and Work

5.0 Introduction

The inauguration of the factory system in the late eighteenth-century British textile industry was a turning-point in the history of waged work; the beginning of a rapid transition from the seasonal rhythms of agricultural toil to the monotonous discipline of industrial time. The 'second industrial revolution' is at one level little more than the fuller and wider extension of this fundamental transition; but such is the period's preoccupation with time that it passes from being a quantitative change, and becomes a qualitatively distinct feature, one moreover with resonances outside manufacturing.

One such resonance is recounted here by Ruth Schwartz Cowan: the application in the 1920s of time-and-motion studies to the unpaid labour of American housewives. The results help confound the conventional view that women's domestic labour was reduced by household technology (shown by Cowan to be a complex of interlocking systems). The role of the housewife certainly changed – from manufacturer to purchaser and transporter – but hours spent on housework if anything increased; and any lessening in the intensity of labour came at a price of increased social isolation. Cowan raises the question of the inevitability of this transformation; she points out that communal means of industrializing domestic labour were technologically available, but were not taken up for over-riding economic and cultural reasons.

A parallel transformation of the labour process took place in manufacturing, and above all in metalworking, where time-and-motion study, and 'scientific management' in general, originated. Between 1870 and 1950, innovations in plant and machinery tended to decrease the physical intensity of manual labour, where they did not actually eliminate jobs; but often the increased control afforded to management led to a stressful 'speed-up' of the pace of work.

In the second reading Daniel Nelson shows how scientific management grew out of Frederick W Taylor's experiments on tool steels at the Midvale Steel Works in Pennsylvania. Taylor's distinctively blinkered attention to production, and authoritarian attitude towards workers, may appear outmoded against the subsequent history of scientific management in the United States; such attitudes would nevertheless have prevailed amongst the Russian managers who took up his ideas in the wake of the Revolution of 1905. Heather Hogan's study of the adoption of scientific management by the St Petersburg metalworking industry adds a new dimension to our understanding of the role of skilled metalworkers in the Russian Revolution, though as her article shows, the rationalization of the labour process was only one ingredient in their militance. Moreover, the adoption of scientific management looks to be less plausibly an epiphenomenon of the wholesale transfer of Western technology, than a response to the peculiar and rapidly shifting socioeconomic context of Russian industry in the immediate pre-revolutionary years.

5.1 Ruth Schwartz Cowan, *More Work For Mother: Technology and Housework in the USA*, 1985

Conventional wisdom has been telling us for many decades that twentieth century technology has radically transformed the American household, by turning it from a unit of production into a unit of consumption. This means that the food and clothing that people once made in their homes is now produced in factories and that what we do 'at home' is consume them (eat the food, wear the clothes) – activities which have economic significance but which require little expenditure of time or energy, only money.[1]

This particular piece of conventional wisdom seems to be subscribed to, ironically enough, by people as diverse as sociology professors and newspaper editors, political conservatives and Marxists. It is a cultural artefact of vast importance, because it has two corollaries that guide people in the conduct of their daily lives: first, that as American families passed from being units of production to being units of consumption, the economic ties that once bound family members so tightly to each other came undone; and second, that as factory production replaced home production, little real work was left for adult women to do at home. Believing that these corollaries are true, many Americans act on this belief in various ways: some hope to re-establish family solidarity by re-learning lost productive crafts such as baking bread or tending a vegetable garden; others dismiss the women's liberation movement as 'simply a bunch of affluent housewives who have nothing better to do with their time'; husbands complain that their wives spend too much time doing inconsequential work ('What **do** you do all day, dear?'); yet housewives can find no reasonable explanation for why they are perpetually exhausted.

The conventional wisdom was once not so conventional. It has its roots in the painstaking sociological observations and patient economic research undertaken by pioneer social scientists.[2] [. . .]

Unfortunately the conventional wisdom may not be as fine a guide to our past or present as we have let ourselves believe.[3] If we adopt the perspective of the historian of technology, we can see that twentieth century household technology consists not of one but of eight interlocking technological systems: the systems which supply us with food, clothing, health care, transportation, water, gas, electricity and petroleum products. Some of these systems, we now know, have followed the conventional model – moving production out of the home and into the factories – but (and this is the crucial point) some of them have not. Indeed, some of the systems cannot even be made to fit this model at all. By tracing them back – in some cases, to the nineteenth century origins

---

Source: Les Levidow and Bob Young (eds), *Science, Technology and the Labour Process: Marxist Studies*, vol. 2, London: Free Association Books, 1985, pp. 88–128.

– we can reveal why the social impact of these technological changes has been a good deal more ambiguous than it at first appeared to be.

From production to consumption

The food, clothing and health care systems are the ones the early social scientists examined in greatest detail; they are the ones, not surprisingly, that best fit the 'production to consumption' model. [. . .]

Less work for mother?

In all three of those technological systems – for food, clothing and health care – the shift from production to consumption occurred slowly, over a long period of time but with increasing momentum, as the nineteenth century turned into the twentieth. Butchering, milling, textile-making and leatherwork had departed from most homes by 1860. Sewing of men's clothing was generally gone by 1880, of women's and children's outerwear by 1900, and finally of almost all clothing for all members of the family by 1920. Preservation of some foodstuffs – most notably peas, corn, tomatoes and peaches – had been industrialized by 1900; the preparation of dairy products such as butter and cheese had become a lost art, even in rural districts, at about the same date. Factory-made biscuits and quick cereals were appearing on many American kitchen tables by 1910, and factory-made bread had become commonplace by 1930. The preparation of drugs and medications had been turned over to factories or to professional pharmacists by 1900. Many other aspects of long-term medical care were institutionalized in hospitals and sanitaria within the next thirty years.

Individual families no doubt differed in the particular times and particular patterns by which they made (or underwent) this transition. [. . .]

Whatever variations of social station or personal inclination there may have been, the general pattern that most American families would follow in most of the arrangements for providing food, clothing, and health care had been settled, at the very latest, by 1930; and this, of course, is the social trend that the earliest social scientists so correctly observed. What they did not observe was the impact that developments in yet a fourth of the household technological systems – the transportation system – would have on the work processes of housework and the time allocations of housewives. As most modern housewives know far too well, you cannot consume frozen TV dinners or acrylic knit sweaters or aspirin or pediatrician's services unless you can get to them, or unless someone is willing to deliver them to your door. In either case you are dependent upon whatever means of transportation is most conveniently at hand. Consequently, the history of urban and rural transportation must also be considered in order to understand why housework did not magically disappear when twentieth century factories, pharmacies and hospitals took over the work that nineteenth century women used to do.

Transportation

The household transportation system developed in a pattern which is precisely the opposite of the food, clothing and health care systems;

households have moved from the net consumption to the net production of transportation services, and housewives have moved from being the receivers of purchased goods to being the transporters of them. During the nineteenth century many household goods and services were delivered to the doorsteps of the people who had purchased them, and many others were offered for sale in retail establishments located a short walk from the houses in which people lived.[4] Pedlars carried pots and pans, linens and medicines to farmhouses and to the halls and stairways of urban tenements. Seamstresses almost always came to the homes of the women and children for whom they were fashioning clothing, and tailors occasionally provided the same service for men. Milk, ice and coal were regularly delivered directly to the kitchens and basements of middle class urban dwellers and not infrequently also into the homes (or at least to the curbsides) of those who were poor. Butchers, greengrocers, coffee merchants and bakers employed delivery boys to take orders from and then carry purchases back to the homes of their more prosperous customers. Smoked, dried and pickled fish, fruits and vegetables, second-hand clothing and linens were routinely sold from pushcarts that lined the curbs and travelled the back alleys of poorer neighbourhoods. Knife sharpeners travelled the streets with flintstones and grindstones on their backs or in their carts; frequently so did the men who repaired shoes and other leathergoods. Bakeries and grocery stores were located in every city neighbourhood, which meant that housewives, children and servants could 'run out' for extra supplies whenever they were needed. Even doctors made house calls.

Under ordinary circumstances the individual urban housewife, whether rich or poor, rarely had to travel very far from her own doorstep in order to have access to the goods and services required for sustenance. For rural householders this was, of course, less true. What shopping there was to be done in rural areas usually waited for the weekly, monthly or (even in some cases) the annual trip into town or the pedlar's arrival.

In the latter decades of the nineteenth century, this pattern of shopping began to change somewhat. In urban areas department stores began to flourish.[5] At first these stores (which, by definition, sold more than one category of goods) were patronised only by 'the carriage trade', people who could afford to keep a horse and carriage and hence could travel to such a store to do their shopping. But later on in the century, their customers changed somewhat as horse-drawn omnibuses, trolley and subway cars made it possible for people of lesser means to take advantage of their centralized locations. Even after the turn of the century, however, the department stores still did not appeal to the poorer segment of the population; their total sales represented only a fraction of the retail sales in most urban areas. Most people living in cities – especially those who were in less comfortable economic circumstances – still acquired most of the goods that they needed, day in and day out, without having to spend very much time either getting to the places in which goods were offered for sale, or in getting their purchases home.

In rural areas, toward the end of the century, the total time spent in shopping for and transporting goods was additionally decreased by the widespread popularity of mail order catalogues.[6] A good part of the business done by urban department stores had always been ordered either by mail (or later, by telephone). In the last decades of the nineteenth century, this service became available to rural Americans as well. Montgomery Ward's and Sears Roebuck (as well as hundreds of smaller enterprises) had entered the mail order business during the 1870s. By the time twenty years had passed, there was virtually nothing – from soup to nuts, from underwear to outerwear, from nails and screws to ploughs and buggies – that rural residents could not have ordered from a catalogue. And clearly they did: the mail order companies were among the country's leading business enterprises by the turn of the century, and they continued to flourish for decades after that. Rural free delivery services, begun on an experimental basis in the 1890s, were extended to most parts of the country before the First World War and so further increased the accessibility of mail ordering. Rural women, like their urban counterparts, simply did not have to spend very much time either in shopping or in transporting the goods they were buying – even if, as the nineteenth century turned into the twentieth, they were buying much more than their mothers and grandmothers had done.

Prior to the advent of the motor car, most transportation services were provided, when they were provided by the household at all, by men or by servants. It was usually the man of the family who hitched the horse to the buggy and went into town to get the mail and buy the flour, cloth or whatever else his family required. Similarly in urban middle class families it was the servant who 'fetched' the doctor, went to market for fresh meat or vegetables, or who drove the family carriage through the streets. Among some immigrant groups, men were responsible for handling the family's money and hence for making the family's purchases in the market place – carrying over Old World traditions into the New; in those families, at least in the first generation of immigration, women were not regular participants in shopping expeditions or decisions. Needless to say, in actual practice typical conditions did not always prevail. There must have been many occasions on which, for one reason or another, mother rather than father made the trip into town, or a middle class urban housewife chose to do her shopping herself; there were certainly immigrant groups (for example the Jews) among whom the women had the gender role of purchasing food and clothing. Yet even granting these exceptions, in the years before about 1920 the shopping and transporting to be done took very little time, and a good part of that time was spent by men.[7]

The motor car

In the years just before and after the First World War, all this began to change. The agent of change was, of course, the motor car. [. . .] People of many social classes were finding that it was possible for them to get to where they wanted to go faster, over longer distances, according to their own schedules – and with carrying capacity attached.

When the automobile began to replace feet and horses as the prime mode of transportation, women began to replace men as household suppliers of the service. For reasons which may only be clear to anthropologists and psychologists, automobile driving was not usually assigned just to members of one sex; in the early days, women were as likely as men to be found behind the wheel. This may have occurred because advertisers of automobiles made a special effort to attract the interest of women, or because the advent of the automobile coincided with the advent of what was then called 'the new girl' – more athletic, better educated, less circumscribed by traditional behaviour patterns. The woman who could drive even became something of a charismatic ideal for a time:

> Like the breeze in its flight, or the passage of light,
> Or swift as the fall of a star.
> She comes and she goes in a nimbus of dust
> A goddess enthroned on a car.
> The maid of the motor, behold her erect
> With muscles as steady as steel.
> Her hand on the lever and always in front
> The girl in the automobile.[8]

The girl who drove an automobile in 1907 (when this jingle was published) was a middle-aged matron by 1930, and by then she was driving not to the moon but to the grocery store. Delivery services of all kinds began to disappear as the nation shifted from an economy dominated by the horse to one dominated by the automobile, and as the Depression created stiff competition among retailers for their share of a declining market. The owners of grocery shops and butchers' markets began to fire their delivery boys in an effort to lower prices and thus more effectively compete with the chain stores and supermarkets which were cropping up throughout the land.[9] Some of the chain stores began to eliminate these services as the Depression deepened. The supermarkets had never offered them; they were, by definition, self-service markets with many departments. After World War II, mail order companies such as Sears Roebuck and Montgomery Ward discovered that they could compete effectively with department stores by opening retail outlets of their own, thus converting their customers into shoppers. Department stores discovered that many of their customers had moved out of the central city neighbourhoods, so they opened suburban branches, which were accessible only by car.[10] Physicians discovered that, without losing a significant number of patients, they could stop making house calls and require that all ambulatory patients (which is itself an ironic euphemism) be brought to their offices.[11] Medical care, in general, became more dependent upon the availability of complex and expensive equipment (X-ray machines, iron lungs, intravenous feedings, anaesthetics). This dependence meant that visiting nurses were less able to cope and hospital visits became more frequent; these required transporting the patient back and forth, rather than waiting at home for the nurse to arrive.[12]

These various individual and corporate decisions were spread out over the course of two decades, but they all pointed in the same direction – to shift the burden of providing transportation services from the seller to the buyer. By the end of the 1930s, the general notion that businesses could offer lower prices by cutting back on customer services was ingrained in the pattern of business relations. The growth of suburban communities in the post-war years did little to alter that pattern: as more and more businesses converted to the 'self-service' concept, more and more households became dependent upon 'her' self to provide the service.

The statistics available about time spent in housework in the years after 1920 reflect this alteration very clearly. Home economists became very interested in time-and-motion studies of housewives during the 1920s, just at the time that proponents of 'scientific management' were investigating the same topics in industry. By the mid-1930s several dozen such studies had been published. Although their methods of investigation varied (as did the size and character of the population being studied), the studies derived fairly consistent figures for the time housewives spent in shopping tasks: on the average, about two hours per week.[13] Roughly thirty years later, when a comprehensive nationwide sample (rigorously standardised) was attempted, the overall figure was eight hours per week – four times greater or roughly the equivalent of one working day out of every week.[14]

Home economists and women's magazine editors who lived through the early years of this transitional period were well aware of what was happening. They believed that the current generation of housewives (of all social classes) was spending increased amounts of time in shopping and that, since their mothers had not been accustomed to such an activity, it was incumbent upon professionals to teach the housewives how to do it wisely. Articles on 'How to Buy Linens' or 'How to Know Good Value When You See It' appeared frequently in women's magazines during the 1920s and 1930s, and home economists began to teach courses in 'consumerism'; they even took children (especially the children of immigrants) on class trips to department stores and hardware shops so as to learn what new goods were available and how to purchase them wisely.[15] The statistics probably underestimate the time that women spent in providing transportation and shopping services. This is due to the way home economists and sociologists categorise time spent in housework; several chores – 'driving a sick child to the doctor', 'waiting in the clinic to see a pediatrician', 'going to the train station to pick up a relative', or 'taking the baseball team to the next town for a game' – get recorded not as 'transportation' but rather as 'family and household management', along with doing the household accounts and repairing the leaking faucets. Clearly, by mid-century the time that housewives had once spent in preserving strawberries and stitching petticoats was being spent in driving or taking the streetcar (tram), shopping, waiting on queues, transporting or accompanying people and goods. Initially this increase in travelling time may have been most apparent in middle class households. However, as 'shopping centres' proliferated across the landscape and neighbourhood shops lost their competitive edge, working class women also discovered that the goods and services that they needed were located further and further away from their homes.

In itself the shift from production to consumption (of such goods as strawberry jam and petticoats) meant a net decrease of time spent in housework, but the shift from the horse and buggy to the automobile meant an increase. The actual physical labour changed markedly – from preserving and sewing and nursing, to driving and shopping and waiting – but the time involved did not necessarily decrease. Of course, preparing a batch of strawberry jam takes more time than picking a jar of it off the shelf in a supermarket. However, 'putting up' jam was a seasonal labour (intense for one or two weeks of the year and non-existent after that), while supermarket shopping is at best a weekly task.

We can speculate endlessly about which form of work is more satisfying or less alienating. Is it better for women to have the 'satisfaction' of preparing jam from scratch or the 'satisfaction' of being out in the social world of the supermarket with other adults? In any case the essential economic and social reality remains.

Nineteenth century women fed, clothed and nursed their families by preparing food, clothing and medication. Twentieth century women feed, clothe and nurse their families by driving, shopping and waiting. The goal is still the same – and so is the necessity for time-consuming work.

Household utilities

The historical development of four household utility systems – those that supply us with water, gas, electricity and petroleum products – reveals other deficiencies in the 'production to consumption' model. This is why I consider their history separately. The ways in which water and energy are supplied to households are important determinants of the labour processes of housework, yet the details of these systems have rarely been taken into account by those who are convinced that women's work has been markedly lightened in the last one hundred years.

Water[16]

[. . .][T]ap water and water closets are only small parts of what we now consider the total water system: the full system contains pipes to conduct water into more than one room of a house, devices for heating some of the water and distributing it, as well as specialized containers for the water (bathtubs, sinks and showerstalls) which make it fairly easy to use the water for different purposes and to dispose of it afterwards. Running household water was available to most residents of American cities by the end of the nineteenth century, but the other parts of the water system were not widely diffused until the early decades of the twentieth. [. . .]

By the end of the 1920s, hot and cold running water had become the norm for American housing, and the architectural form of the modern bathroom had solidified.[17] Most urban and small town middle class families had hot and cold water taps in their kitchens and at least one separate room had become a bathroom, equipped with two sets of taps (for sink and tub), ceramic tiles (made by the manufacturers of sinks and tubs) and water closets. The water was likely to be available at any hour of the day or night at sufficient pressure to make all those fixtures function. For rural families and for the urban poor, the introduction of water systems was somewhat delayed, but not by much. Partly as a result of municipal building codes and partly as a result of rural electrification, many such families achieved access to the full water system during the decade of the 1930s.[18] [. . .]

Women shared in the general lightening of labour that occurred when taps replaced pumps and hot water heaters replaced kettles, but the advent of indoor plumbing created new tasks that had never existed before. Because of the way households were organised at that time, these tasks quite 'naturally' fell to women. One such task was cleaning the bathroom; this was not light work and, given the frequency with which the room was used and the seriousness of the diseases that were thought to result from unsanitary conditions, it was not casual work either. By the 1920s and 1930s most men were employed (when employment was available) outside their homes, and most children spent most of their youth in school. Household cleaning, in any event, had traditionally been designated as 'women's work'. Thus it was hardly surprising that 'cleaning the bathroom' was added to the list of mother's chores as soon as a bathroom was added to a house or apartment. Indeed the advent of indoor plumbing may be another good example of modernization entailing a shift from men's work to women's work, for there is some evidence to suggest that men had been commonly assigned to building, maintaining and cleaning outdoor privies.

In any event bathrooms were not the only things that women began to keep scrupulously clean as soon as indoor plumbing made scrupulousness a widespread possibility. In the period between the two world wars, sustained efforts were made to convince American housewives that household cleanliness was the best way to improve the health and the social status of their families.[19] Home economists, social workers, rural extension agents, columnists in women's magazines, manufacturers of soaps, cleansers and deodorants, nurses, advertising agents, physicians and public health authorities all participated in these campaigns, many (but not all) of which were directed at housewives who were either recent immigrants or poor or both. Some social campaigns fell upon deaf ears (as, for example, the campaign to get young people to stop smoking) but this one did not. The germ theory of disease had just begun to permeate the public consciousness; the memory of devastating epidemic diseases (influenza, cholera, typhoid) was very fresh; the infant mortality rate was still terrifyingly high; and dirty homes, dirty clothes and dirty bodies still seemed (as they always had seemed) to be the single most obvious indicator of poverty and the single most obvious impediment to acceptance in the world of 'white collar' employment. Thus it is hardly surprising that between the two world wars American housewives began to attend to the cleanliness of bathrooms and kitchens, underwear and shirtcollars, teeth, breath, and underarms, sheets, towels and handkerchiefs with fervour that now appears alternately humorous and horrifying to their children and grandchildren. Many commentators (including this author, in an earlier essay) have written about this obsession with cleanliness as if housewives were conned into an ideology which only served the interests of the property-owning classes (particularly those property-owners who were manufacturers of soaps and plumbing supplies.)[20] This may have been true, but it was not the whole story, for obsessions with cleanliness served the interests of the housewives as well, or – at the very least – the housewives had every good reason to believe that they did: epidemic diseases were brought under control; the infant mortality rate did go down; and many sons and daughters did succeed in obtaining white collar jobs.

However, what did these improvements entail for women's domestic labour? If it makes sense to say that, without modern water supply systems, women (and other members of the family) had to 'produce' water for cooking, and hot water for bathing and laundering, then it also makes sense to say that *with* such systems women have to 'produce' clean toilets, bathtubs and sinks. The item produced has changed, but the labour has not. Indeed, as we all know, standards of personal and household cleanliness have increased markedly during the twentieth century: sheets, underclothes, table linens are changed more frequently, while floors, carpets, fixtures are kept freer of dust and grime. Or, to put it somewhat more accurately: more people are becoming accustomed to live at standards of cleanliness that were once possible only for those who were very rich. 'Increased standards of cleanliness', when translated into the language of production and consumption, essentially means 'increased productivity'. Women have always been responsible for keeping their families and their homes clean; with the advent of modern water (and other utility) systems, they became responsible for keeping them even cleaner. Given the gender-role division of labour in the early twentieth century, this work fell to women. Hot and cold running water in the kitchen and the bath meant that men and children were increasingly free to labour in other places and that women were able to become more productive at home, whether or not they laboured in other places as well.

Gas

The other household utility systems had a similar impact on the work processes of housework, in part because they reached a stage of fairly complete development – at least in the USA – at about the same time as the water supply system. [. . .]

At the turn of the twentieth century, the gas companies became aware that their lighting business was eroding ever more swiftly and was unlikely to return.[21] Consequently, they began an active campaign to promote gas for heating; partly as a result of their efforts, the transition was achieved rather swiftly. In 1899 roughly 75% of the gas being produced nationwide was for illumination; the largest part of the remainder was consumed in industrial processes. Twenty years later, in 1919, only 21% was still used for this purpose, while 54% was consumed as domestic fuel.[22] The trade association of gas companies, the American Gas Association, subsidised development work on the improvement of gas cookers or stoves, hot water heaters and hot air furnaces; this tactic proved effective. In 1915, for example, the American Stove Company brought out the first effective oven thermostat; at about the same time, stove manufacturers began to shift from cast iron to enamelled surfaces.[23] The gas companies also began to offer reduced rates to domestic customers who purchased heating apparatus – a useful technique pioneered, ironically, by the electric companies. Some companies also began to convert from manufactured to natural gas (a mixture of organic gases located in underground deposits which is considerably more efficient and less toxic), further lowering the cost of gas heating and making it more attractive to homeowners. By 1930s gas cooking prevailed over all other forms in the United States: almost 14 million households cooked with gas, while only 7.7 million still used the older fuels (coal and wood), 6.4 million used oil, and 0.9 million used electricity.[24] By 1930 gas space and water heating, although not as prevalent as gas cooking, were also becoming commonplace.

Electricity

Electricity and oil, as competitors with gas, were equally a part of that modernization process. [. . .]

The a.c. motor was able to invade the household, only because its price and the price of electricity was made low enough to be affordable by householders of moderate means. This was accomplished earlier in the USA than in other countries which had started on the road to electric development, because standardization occurred much earlier in the USA.[25] So few American companies held the crucial patents on the manufacture of electrical goods that it was a relatively simple matter for them to agree on standardization, and then to ensure standardization by cross-licensing their patents. Fairly early on, the electrical utility companies formed themselves into a national association, the National Electric Light Association, as did the electrical manufacturers, the National Electrical Manufacturers Association. This also helped to insure the spread of standardization. The federal government did its part too, not so much by direct action as by encouraging cross-licensing as a technique to avoid charges of monopolistic control. The end result was that by 1910 the entire country had been standardized on one form of electric current: alternating-current, generated at 60 cycles, transmitted at high voltages, but stepped down by transformers to 120 volts before entering households. Since electrical manufacturers could anticipate a nationwide market for products made to the same standard, mass production techniques became a worthwhile investment for them. The end result for Americans was that the price of electric energy and electric appliances fell markedly between 1900 and 1930 and remained (during all that period and for many years thereafter) markedly lower than similar prices abroad. [. . .]

By 1900 small a.c. motors were being sold with the intention that householders would connect them to the foot-treadles of sewing machines or the hand-cranks of washing machines. A decade later appliances based upon the electric resistance coil (toasters, irons, hot water urns, and hair curling devices) were becoming popular, since the substitution of nickel-chromium alloys for platinum in those coils has made them simultaneously less expensive and more reliable (as the platinum coils had burned out fairly easily). Between 1910 and 1920 all-of-a-piece electric sewing and washing machines came on the market, as did the early forms of the electric vacuum cleaner – essentially an electric fan hooked up to a carpet sweeper. Electric refrigerators and dishwashers were also available by the time the USA entered World War I, but they were little more than toys for the very rich, as they tended to be both extremely expensive and extremely unreliable. During the 1920s the price of washing machines, sewing machines and vacuum cleaners fell (as mass production techniques were introduced) and the reliability of the refrigerator increased as new designs crowded each other on the market. Electric cake mixers (essentially old-fashioned egg beaters attached to electric motors) and food grinders also became popular during this period. By 1930 electric stoves were beginning to compete effectively with gas ranges, because electromechanical thermostats had been introduced during the twenties. These thermostats also increased the popularity of electric toasters, irons, hot water and space heaters. During the 1930s the price of electric refrigerators plunged, partly because of the introduction of mass production in this section of the industry, and also partly because of stiff competition for sales during the Depression.

At the end of the 1930s, the automatic washing machine was introduced. Since this machine could spin dry as well as agitate, the time-consuming chore of running wet clothing through wringers between each part of the cycle was eliminated. Although this washing machine (along with the electric dishwasher and the clothes dryer) did not become widespread until after World War II, electric service and electric appliances had become sufficiently widespread among different classes of the population for us to say that, along with gas appliances, they had begun to co-operate in the process or reorganizing the structure of household work by the outbreak of World War II. [. . .]

Petroleum products

Oil and its associated appliances were introduced into American homes somewhat later than gas but earlier than electricity.[26] Kerosene lamps had three advantages over gas lights: they were portable, they provided a pleasanter and steadier light, and they could be used in households that had no access to gas service – which, in the last quarter of the nineteenth century, still meant the majority of the nation's households. As the oil companies further developed their skill in bringing kerosene to market, they also managed to lower its price, so that kerosene lamps acquired a fourth advantage – economy. In the last quarter of the century, kerosene was still competition for gas.

By the turn of the century, however, the petroleum refiners had come to realize – as had the gas companies – that the handwriting was on (or rather 'in') the wall: electricity was going to erode the market for kerosene fairly quickly. As the refiners already had well-established distribution networks keyed to the residential customer, they were very reluctant to surrender these; hence they began to take defensive action. Kerosene and gasoline stoves had been on the market for several decades, but now the refineries began to give them away free or at much reduced prices, in order to stimulate demand for their fuel. They also began to encourage furnace manufacturers to develop designs that would enable homeowners to convert from coal to oil for central heating. These 'oil burners' first went on the market during the 1920s but they did not begin to make serious inroads on the domestic heating market until after World War II. At that time the price of home heating oil fell relative to the price of coal, and a strike wave in the coal fields made the ready availability of that product somewhat uncertain.

Central heating, with all that it entails both for comfort and for convenience, had become an American norm even before oil became the principal fuel used for this purpose.[27] Large coal furnaces for household use had come on the market during the second half of the nineteenth century; these came complete with plumbing systems (if they were to be used to provide heat by piping hot water or steam through a house) and venting systems (if they were to be used to provide hot air). By the turn of the century, certain automatic features had been incorporated in these furnaces, so that, for example, they did not have to be continually resupplied with coal during the day, but complete thermostatic controls were not available until the 1920s. By that time, gas furnaces with similar features were also available and competitively priced; in some cities 'central heating' was available as a municipal service. Steam was generated in large central plants and piped, underground, to individual homes and apartment buildings for a monthly fee.

Installation of central heating was always an extremely expensive undertaking. The furnace itself was a major investment; during the last decades of the nineteenth century, for example, a furnace cost as much as a teacher's annual wage. So were the labour costs for installation of pipes, radiators and vents. In addition, owners of older homes often had to pay to have a special basement room dug and cemented for the furnace, since walk-in basements had not been standard features of nineteenth century residences. Consequently central heating remained a comfort enjoyed only by those who were fairly rich until after World War I, when – partly as a result of the boom in construction of new homes for the middle classes – it began to spread further down the economic scale. The process was slow, however, since central heating remained the most expensive of all improvements that could be made to an older home (or apartment building) and since the poor lived in the oldest segment of the housing stock. [. . .]

Central heating was not available to those in straitened economic circumstances until the years after World War II. It was a national norm for housing in the sense that those who did not have it were regarded as seriously disadvantaged, but not in the sense that the bulk of the population was yet able to receive its comforts and its bills.

Nevertheless, with the exception of central heating and some advanced electrical appliances, most Americans made the transition to modern fuels before the outbreak of World War II. The symbolic nineteenth century kitchen, dominated by the great coal or woodburning stove, was transformed into the symbolic twentieth century kitchen dominated by the gas, oil or electric range. Hand tools were replaced by power tools, and human energy by chemical and mechanical energy. Scrubbing boards were discarded and power washing machines took their places; carpet beaters and sweepers were thrown away and vacuum cleaners acquired. Gas lights and kerosene lamps were wired and fitted with electric bulbs. Coal and wood no longer had to be hauled into the kitchen and ashes no longer had to be hauled out of it. Cast iron stoves no longer had to be rubbed with stoveblack; kitchens and parlours were no longer plagued by a constant residue of greasy ash. Light and heat could now be obtained, at almost any hour of the day or night, without very much attention to the maintenance of the appliance or the supply of fuel. Most people seemed to have acquired conveniences and appliances as soon as they were in a position to pay for them and, given the prevalence of instalment credit plans after 1918, sometimes even sooner than that.

Mother's helpers: labour-saving utilities?

The multitudinous chores that constitute housework were, quite obviously, totally reorganized by these technological changes. Yet – as with the water supply system – the impact of those changes on the nature and amount of work that housewives had to do was mixed. Consider, for a moment, the three parts of the work process of cooking which are directly connected to the appliance which we call a stove or cooker: fuel must be supplied, pots must be tended and, subsequently, the appliance must be cleaned. Modernization of the fuel supply systems for stoves eliminated only one of those steps (the first); and *that* was the step which, in the coal and wood economy, was most likely to be assigned to the men and boys in the family. Or consider the process of doing laundry: washing sheets with an automatic washing machine is considerably easier than doing it with one that has a wringer, in turn considerably easier than washing sheets with a scrubboard and tub.

Yet the easiest solution of all (at least from the housewife's viewpoint) is to have someone else to do it altogether. This was a very common practice (even in fairly poor households) in the nineteenth century and even in the first few decades of the twentieth century.[28] Laundresses were the most numerous of all specialized house servants; many women who did their own cooking, sewing and housecleaning would have a laundress 'in' to do the wash, or failing that, would send some of it 'out' to be done. 'Out' might have been the home of the laundress herself, or – especially in the early decades of the twentieth century – it might have been a commercial laundry, business establishments that were especially numerous in urban and suburban communities.[29] The advent of the electrically powered washing machines (as well as the advent of synthetic washable fabrics) coincided with the advent of 'do-it-yourself' laundry; this meant that the woman endowed with a Bendix would have found it *easier* to do her laundry, but, simultaneously, would have done *more* laundry, and more of it herself, than either her mother or her grandmother had.[30]

We could multiply examples endlessly. Yet these two should suffice to reveal the reasons why modern fuel supply systems did not necessarily lighten the work of individual housewives, though they certainly did reorganize it. Some of the work that was eliminated by modernization was work that men and children – not women – had previously done: carrying coal, carrying water, chopping wood, removing ashes, stoking furnaces, cleaning lamps, beating carpets. In the homes of the poorer people, some of the work was made easier but its volume increased: sheets and underwear were changed more frequently, so there was more laundry to be done; diets became more varied, so there was more complex cooking to undertake; families moved to larger quarters, so there were more surfaces to be cleaned. In the homes of the more affluent, much of the work formerly done by hand had been done by servants, so that some of that work now came to be done by the housewife herself, aided by machines; indeed, many middle class families purchased appliances precisely so that they could dispense with servants. From 1900 to 1950 the ratio of servants to households fell from 1 per 15 to only 1 per 42 – a decrease which coincided with an equally marked proliferation of washing machines, dishwashers, vacuum cleaners and refrigerators.[31] Finally, the housewife became burdened with some of the work which had previously been allocated to commercial agencies – laundry, rug cleaning, drapery cleaning, floor polishing – as new appliances were invented to make the work feasible for the average housewife and as the costs of labour (except, of course, of the housewife's labour) escalated in the post-war years.

Haven in a hearthless world: The production and reproduction of real life

The changes that occurred in the labour processes of housework during the twentieth century were not the ones that the conventional sociological model predicts. American households and American housewives did not shift from production to consumption, but rather from one level of production to another. Prior to the advent of industrialization – which in the USA means prior to 1860 – American households, and the adult women who lived within them, produced goods intended for sale in the market place, but they also produced goods and services that were intended for use at home: foodstuffs, clothing, medicines, meals, laundry, health care and much more. After industrialization began (that is, after roughly 1860), households stopped producing goods for sale; but that did not mean – and this is the crucial point – that they ceased to be productive locales, because they continued to produce goods and services intended for use at home or – in Marxist terms – intended for the production and reproduction of labour power. When the household itself began the process of industrialization (that is, after roughly 1920), those productive functions did not leave the home; the industrialization of the household did not entail, as the industrialization of the market had, the centralization of productive processes; the household continued to be the locale in which meals, clean laundry, healthy children and well fed adults were 'produced', and housewives continued to be the workers who were principally responsible. What changed most markedly was the productivity of the workers. Modern technology enable middle class American housewives of 1950 to produce singlehandedly what their counterparts of 1850 had needed a staff of three or four to produce. It also enabled many (though not all) working class housewives to achieve levels of household health and cleanliness that had been totally beyond the reach of their mothers and grandmothers.

While this was happening, the daily round of household activities – the work processes of housework themselves – changed drastically. Housewives ceased to be manufacturers and instead became purchasers and transporters. The housewife of 1950 probably did not know how to preserve strawberries or bake bread (or if she did know how, she was not putting her knowledge to use) but she did know how to shop for bargains, drive a car, and get her children to doctors, dentists, music lessons, birthday parties and baseball games at the proper times and places. Housewives also stopped working in tandem with their husbands and their children; as housework became more productive it also became more socially isolated. Cooking no longer required the assistance of men (to carry the coal, stoke the fires) or children (to carry out the ashes and clean the stove); laundry no longer required additional help (for carrying the hot water, or providing extra hands for the wringing, or hauling the wet wash to the line to dry); even caring for the ill no longer required two or three adults to do shifts by the bedside of a sick child or to manage the activities of those who were not sick. Men began to work further and further away from their homes and to spend many hours commuting; children stayed in school longer and left home earlier when their schooling was finished; grandparents had occupations and households of their own to care for; maiden aunts took paid employment outside the home; servants became increasingly difficult to find and increasingly difficult to afford. The housewife changed from being the supervisor and manager of other people's labour (in addition to her own), to being the sole worker in a decentralized manufacturing plant. At the same time, because modern technology eliminated some backbreaking chores, her labour became less exhausting – freeing her to become even more productive in the time that was available.

These alterations in the labour process of housework were not accidental; they occurred for reasons which are complex and difficult to describe, but which are nonetheless ascertainable. Certainly there were alternatives. Housework could have been industrialized in the same mode that market work was – in a centralized mode; the equipment and the social organization for communalizing housework existed in the early decades of the century.[32] Co-operative laundries, communal kitchens, neighbourhood bathhouses, even communal nurseries and kindergartens dotted the landscape, from the tenements of lower Manhattan to the rich suburbs of Chicago, to the sparsely settled rural hinterlands of Minnesota and Oregon. Indeed, in the early 1920s several social commentators feared (or hoped, depending on their political persuasions) that the country was indeed going down the road of socialism of the most fundamental sort – that is, the socialism of the home.

Yet, having started down the road, Americans did not follow it all the way. By the end of the Great Depression, virtually all these cooperative experiments had been abandoned, and with them the potential for large-scale domestic technologies. That potential was stymied for many reasons. Its realization would have required radically altering existing housing stock. Large corporations, such as General Electric and Westinghouse, were quick to understand the economic benefits that would result if they switched from 'producer's goods', to 'consumer's goods'. In addition – especially during the Depression – there were few opportunities for gainful employment for adult women in the labour market and many social reasons why people preferred to have wives and mothers remain at home. Much of (what once had been) men's work in the home was eliminated by the earliest spate of inventions, thus increasingly releasing men from the necessity of being present at home. Finally, and perhaps most crucially, people (both men and women) were reluctant to give up those symbolic activities which are so intimately connected with the emotive and affective bases of family life – particularly the activities of eating, cooking, cleaning and childcare. As a group, Americans simply did not want to give up to the market place and to the community (to 'strangers') that group of activities which they regard as 'private' and symbolically meaningful.

The productive work which is still being done in American homes is difficult to 'see' because the reigning theory of family history tells us that it should not be there, and the reigning methodology of social sciences cannot be applied to it. Household work is not counted as productive labour because, at least in part, economists and sociologists cannot measure it. Of course, they can fairly easily quantify what people are consuming: how many cans of peas? how many dollars worth of stockings? But they cannot place a dollar value (to choose a simple example) on a nutritious meal. Nor can they begin even to estimate how many such meals are prepared in households throughout the year, partly because the workers who prepare them are neither paid nor timed. But people demonstrate through their behaviour that they know that the labour is there – and that it is economically valuable labour. Housewives still fall into bed exhausted at night. They know perfectly well that – whether they call their activity 'consumption', 'purchasing' or 'maintaining our social status' – it still takes time and lots of energy. Men still marry and, when divorced, marry again; men behave as if they know (leaving aside considerations of companionship, sexuality and affection) that women's labour is difficult to live without – modern technologies notwithstanding.

Notes

This essay is adapted from Ruth Schwartz Cowan, *More Work for Mother: The Ironies of Household Technology from the Open Hearth to the Microwave* (New York: Basic Books, 1983).

1 This conventional wisdom was articulated in numerous sociology textbooks of the 1950s and 60s. For example: Ernest W. Burgess and Harvey J. Locke, *The Family* (New York, 1950; or Paul R Horton and Chester L. Hunt, *Sociology* (New York, 1964). Recent Marxist histories of the family such as Eli Zaretsky, *Capitalism, The Family, and Personal Life* (New York, 1977) also adopt it.

2 Helen and Robert Lynd, *Middletown: A Study in American Culture* (New York, 1929); Simon Patten, *The Reconstruction of Economic Theory* (Philadelphia, 1912); Thorstein Veblen, *The Theory of the Leisure Class* (New York, 1899); William Fielding Ogburn and Meyer Nimkoff, *Technology and the Changing Family* (Cambridge, Mass., 1955); Margaret Gilpin Reid, *Economics of Household Production* (New York, 1934); Hazel Kyrk, *A Theory of Consumption* (Boston, 1923).

3 For criticisms arising from history of technology see Ruth Schwartz Cowan, 'The "Industrial Revolution" in the Home: Household Technology and Social Change in the Twentieth Century', *Technology and Culture*, 17 (January, 1976) 1–42; from social history see Elizabeth H. Pleck, 'Two Worlds in One: Work and Family', *J. of Social History*, 10 (Winter, 1976) 178–195. Within sociology itself the model has not gone unassailed: For example see William J. Goode, *After Divorce* (New York, 1956) esp. Ch. 1.

4 I have been unable to locate a comprehensive survey of the history of transportation and shopping patterns in the United States. The best approximation is a combination of George Rogers Taylor, *The Transportation Revolution* (New York, 1951) and Daniel Boorstin, *The Americans: The*

Democratic Experience (New York, 1973) Chs. 9–14.

5 On the development of the department store see Harold Pasadermadjian, *The Department Store: Its Origins, Evolution and Economics* (Chicago, 1954) and Ralph M. Hrant, *History of Macy's of New York 1859–1919* (Cambridge, Mass., 1943).

6 On rural mail order shopping see Boris Emmet and John E. Jeuck, *Catalogues and Counters: A History of Sears, Roebuck and Company* (Chicago, 1950); and Wayne E. Fuller, *R.F.D., The Changing Face of Rural America* (Bloomington, 1964).

7 There is no single secondary source from which the judgements made in this paragraph derive. My assessment of the rural sexual division of labour in transportation is based upon my reading of diaries and letters too numerous to list here; one good example would be, *The Diary of Elizabeth Koren*, translated and edited by David T. Nelson (Northfield, Minn. 1955) which covers the years from 1853–1855. Diaries and letters are also the source of my judgement that, for the urban middle classes, household servants did much of the fetching and shopping that needed to be done by the household. On the shopping arrangements of immigrants, see for example Charlotte Baum, Paula Hyman and Sonya Michel, *The Jewish Woman in America* (New York, 1976) Ch. 3 for Jews, and Virginia Yans McLaughlin, *As the Fingers of One Hand* (New Haven, 1979) p. 6 for Italians.

8 The poem comes from *Leslie's Illustrated Weekly* in 1907 as cited in Rudolf E. Anderson, *The Story of the American Automobile* (Washington, D.C., 1950) p. 198. The 'new girl' is described in Margaret Deland, 'The Change in the Feminine Ideal', *Atlantic Monthly*, 105 (1910) 290–295.

9 Rom Markin, *The Supermarket: An Analysis of Growth, Development and Change* (Pullman, Washington, 1968) Ch. 1.

10 Pasadermadjian, *The Department Store*, Ch. 5.

11 Rosemary Stevens, *American Medicine and the Public Interest* (New Haven, Connecticut, 1971), p. 145 reports that a survey of private practitioners in Philadelphia in 1929 revealed that physicians were spending roughly six hours a week making house calls during an average working week of 50 hours; the rest of the time they were in offices, hospitals and clinics. On the same phenomenon in rural areas see Reynold M. Wik, *Henry Ford and Grass Roots America* (Ann Arbor, Michigan, 1972), p. 23; and Michael L. Berger, *The Devil Wagon in God's Country: the Automobile and Social Change in Rural America, 1893–1929* (Hamden, Conn., 1979), Ch. 7.

12 *Medical Care for the American People, The Final Report of the Committee on the Costs of Medical Care* (Chicago, 1932), Ch. 1.

13 Time studies of housewives done during the period 1920–1970 have been described and analysed in Joann Vanek, 'Keeping Busy: Time Spent in Housework, United States, 1920–1970', (unpublished dissertation, University of Michigan, 1973). A very brief published version of this, which gives charts of changes in time spent in shopping, is Joann Vanek, 'Time Spent in Housework', *Scientific American*, 231 (November, 1974) 116–121. Summaries of the data to the early 1930s can be found in Benjamin Andrews, *The Economics of the Household* (Chicago, 1936) pp. 442–443; and *Household Management and Kitchens*, volume IX of the President's Conference on Home Building and Home Ownership (Washington, D.C. 1932) pp. 65–68.

14 James N. Morgan, I.A. Sirageldin and N. Baerwaldt, *Productive Americans* (Ann Arbor, 1966) pp. 111–112.

15 For examples of articles in women's magazines, see 'How to Buy Towels', *Ladies Home Journal, 45* (February, 1928) 134; or 'Buying Table Linen' *Ladies Home Journal, 45* (March, 1928) 43. For home economists' activities

see hundreds of articles in the *Journal of Home Economics*, 1920–1930 as well as Robert and Helen Lynd, *Middletown*, p. 196.

16 There is no single history of water supply systems in the United States since most authors treat each part of the system separately (sewers and aqueducts, for example, separately from plumbing and bathroom fixtures). The account which follows is based upon Nelson M. Blake, *Water for the Cities: A History of the Urban Water Supply Problem in the United States* (Syracuse, New York, 1956); Ellis L. Armstrong *et al.*, eds., *History of Public Works in the United States, 1776–1976* (Chicago, 1976) Chs. 8 and 12; Lawrence Wright, *Clean and Decent: The Fascinating History of the Bathroom and the Water Closet* (New York, 1960); Elizabeth Mickle Bacon, 'The Growth of Household Conveniences in the United States, 1860–1900' (unpublished dissertation, Radcliffe College, 1942) *passim*; Siegfried Giedion, *Mechanisation Takes Command* (New York, 1948) Pt. VII; Clark, *A History of Manufactures in the United States*, various chapters in volumes II and III which consider 'sanitary pottery' and 'pipes and tubes'; *Historical Statistics of the United States* (Washington, D.C., 1960) Series D327, p. 76.

17 For descriptions of this bathroom see, e.g.: Helen M. Sprackling, 'The Modern Bathroom', *Parent's Magazine* (February, 1933) 25; or any issue of such a trade journal as *Merchant Plumber and Fitter* during the 1920s.

18 On patterns of diffusion of water supply and plumbing facilities see Robert and Helen Lynd, *Middletown*, esp. p. 97; Robert and Helen Lynd *Middletown in Transition* (New York, 1937) pp. 557–558; Lenz Houghteling, *The Income and Standard of Living of Unskilled Labourers in Chicago* (Chicago, 1927) esp. p. 109; *Zanesville, Ohio and Thirty Six Other American Cities* (New York, 1927) *passim*; United States Department of Agriculture, *Farm Housing Survey* (Washington, D.C., 1934) *passim*; Faith M. Wilson and Alice C. Hanson, *Money Disbursements of Wage Earners and Clerical Workers in Eight Cities*, 1934–1936, U.S. Bureau of Labour Statistics, Bulletin 636 (Washington, D.C. 1940) p. 52.

19 A partial view of the extent to which personal cleanliness, personal health and public health were linked together can be derived from George Rosen, *Preventive Medicine in the United States, 1900–1975* (New York, 1975). See also the advertising campaign of the Cleanliness Institute, 'Self respect thrives on soap and water,' which ran in American popular magazines, such as *The Ladies Home Journal*, during the second half of the 1920s.

20 Cowan, 'The Industrial Revolution in the Home . . .'; and, as another example, Barbara Ehrenreich and Dierdre English, 'The Manufacture of Housework', *Socialist Revolution*, 26 (October–December, 1975) 5–40.

21 On the period in which the gas, electric and petroleum industries competed against each other for lighting and heating business see: Harold Williamson and Arnold R. Daum, *The American Petroleum Industry: the Age of Illumination, 1859–1899* (Evanston, Illinois, 1959), Chs. 13, 20 and 25.

22 Harold F. Williamson *et al., The American Petroleum Industry, 1899–1959: The Age of Energy* (Evanston, Illinois, 1963) p. 39.

23 Jane Busch, 'Cooking Competition: Technology on the Domestic Market', *Technology and Culture* 24 (April 1983) 222–244.

24 'Our Market for Ranges – and Our Competition', *Electrical Merchandising*, 43 (May, 1930) 38, as cited in Busch, 'Cooking Competition'.

25 On standardization and its effect on the price of electrical [appliances], see Harold C. Passer, *The Electrical Manufacturers, 1875–1900* (Cambridge, Mass., 1953), pp. 276–34; and Frank Joseph Kottke, *Electrical Technology and the Public Interest* (Washington, D.C., 1944).

26 Williamson and Daum, *The American Petroleum Industry: The Age of Illumination*, and Williamson *et. al., The American Petroleum Industry: The Age of Energy*, are the basic sources of the discussion which follows.

27 On the history of central heating devices see: Bacon, 'The Growth of Household

Conveniences . . .', *passim*; Eugene Ferguson, 'Victorian Heating Systems and Kitchen Equipment', (unpublished paper, 1974); Williamson, *et. al., The American Petroleum Industry*, pp. 265 and 293; and Lawrence Wright, *Home Fires Burning: The History of Domestic Heating and Cooling* (London, 1964). For the 20th century the best sources of information are trade journals such as *Merchant Plumber and Fitter, Gas Age*, and *Heating and Refrigeration News.* [. . .]

28 My judgement about the frequent assignment of laundry work to servants is based upon my reading of diaries and letters (see n7), and household manuals, e.g., Christine Frederick, *Household Engineering* (New York, 1920) esp. pp. 70–77. For an example of the practice of hiring laundresses even in very poor households see Margaret J. Hagood, *Mothers of the South* (New York, 1942) p. 67. On the numerical preponderance of laundresses amongst household servants see David M. Katzman, *Seven Days a Week: Women and Domestic Service in Industrialising America* (New York, 1978) pp. 46–47.

29 On the growth of the commercial laundries in the twentieth century see: Heidi I. Hartmann, 'Capitalism and Women's Work in the Home, 1900–1930', (unpublished Ph.D. dissertation, Yale University, 1974) Ch. 6.; and Frank D'Armand, *Commercial Laundries* (New York, 1954).

30 For documentation of the increase in time spent in laundry work and quantity of laundry done see: Maud Wilson, 'Laundry Time Costs', *J. Home Economics*, 25 (1930) 735–738; and Amy Hewes, 'Electrical Appliances in the Home', *Social Forces*, 2 (December 1939) 235–242.

31 These are my figures, computed from Historical Statistics, Series A255 ('number of households') p. 16 and Series D84 ('private household workers') p. 74.

32 The discussion which follows is based upon Dolores Hayden, *The Grand Domestic Revolution: A History of Feminist Designs for American Homes, Neighborhoods, and Cities* (Cambridge, Mass., 1981).

5.2 Daniel Nelson, *Frederick W. Taylor and the Genesis of Scientific Management*, 1980

Taylor's [metal-cutting] experiments began between late 1880 and late 1881, very likely at the former date. In either case he was far in advance of his contemporaries. [. . .]

If the experiments did not immediately provide the answers that Taylor sought, they did produce (in addition to the inventions) new information about the relationships between the 'speed' or rate at which the metal in a lathe revolved, the 'feed' or application of the tool, the depth of the cut, the power required for various processes, and external factors, such as the use of water to cool the cutting tool. [. . .]

There were three other results of the metal-cutting experiments that Taylor noted in 1906: in 1883, 'the starting of a set of experiments on belting;' in 1884, 'the construction of a tool room for storing and issuing tools ready ground to the men;' and in the period 1885 to 1889, 'the making of a series of practical tables for a number of machines . . . [by] which it was possible to give definite tasks each day to the machinists who were running machines.'[1] These developments were logical

Source: Daniel Nelson, *Frederick W. Taylor and the Rise of Scientific Management*, Madison: University of Wisconsin Press, 1980, pp. 38–45.

outgrowths of Taylor's technical work. They showed the many possible ramifications of the metal-cutting investigation. Equally important, they suggested the close ties between Taylor's technical virtuosity and his solutions to the apparent deficiencies of the factory.

The Genesis of Scientific Management

Despite his 'wonderful and great,' and presumably costly, projects, Taylor considered himself primarily a businessman and acted accordingly. As gang boss and machine shop foreman he had adopted policies that benefited the company; as master mechanic and chief engineer he worked assiduously to advance his employer's interests. Even his inventive activities were usually compatible with his role as junior executive and apprentice shop superintendent; he undertook the metal-cutting experiments to improve the operation of the shop, as well as to satisfy his curiosity. It is not surprising that Taylor soon began to experiment with the organization, as well as the machinery, at Midvale, or that he attacked problems that first became apparent as a result of his day-to-day duties. By the mid-1880's he had independently duplicated important features of the systematic management plans of Towne and others. In the process he had also devised an approach to the incentive wage that would ultimately make him the most famous – and controversial – engineer of his time.

As a manager Taylor adopted the authoritarian style characteristic of late-nineteenth-century executives. Charles W. Shartle, who worked as a repairman at Midvale from 1884 to 1887, recalled how Taylor, like Brinley and Davenport, would not tolerate excuses or imperfect work: 'When he would give us a job he would tell us when he expected it finished, and there was no reason to ask questions about it, or argue the matter, and it was up to us to finish it even if we had to work all night to get it done.'[2] Taylor was equally insistent that his men follow orders. Yet like many contemporary executives, he combined his authoritarianism with a friendly and sympathetic personal manner. He spoke to the men in their own language, tempered his criticisms with 'a touch of human nature and feeling,' and encouraged the workers to discuss their problems with him. He adopted this approach in part as a hedge against unrest and unions. From all accounts, he was personally popular. Despite widespread resentment of his fining system, most Midvale workers respected him and his work.

By 1883 Taylor's critique of the plant management, demands for cost reductions, and most of all the dynamics of the metal-cutting experiments led him to broaden his efficiency campaign. If machine performance could be improved by careful study, reorganization, and innovation, he reasoned, could not the quality of the management and the workers' efforts be improved by a similar process?

Unfortunately, the precise chronology of Taylor's administrative reforms is not clear from his writings and papers, and his rapid advancement from machine shop foreman to chief engineer in 1883 and 1884 makes it virtually impossible to reconstruct his work during those years. Apparently he hired additional clerks, devised a system of production controls, and introduced stopwatch time study and an incentive wage, the 'differential piece rate,' before 1885. His greatest impact was in the machine shop, but he reorganized other departments as well. Though crude by later standards, Taylor's plans were innovative, perhaps even unique. In any case they established Taylor as a leader of the systematic management movement.

Taylor's most ambitious initiative was a complicated production control system for coordinating the work of the functional departments. The 'chief idea,' he explained in 1886, was that 'authority for doing all kinds of work should proceed from one central office.' In each department a clerk controlled the flow of orders and records between the central office, the supervisors, and the workers. 'The clerk or clerks . . . write . . . under the direction of the foremen . . . orders stating that work is to be done; what order number to charge it to, and what drawings and tools are to be used, etc.' The clerk posted the orders on a bulletin board and sent each worker 'a note or card . . . which refers him to the more elaborate order.' When the worker completed the job, he or the foreman entered 'all desired information' on the note and returned it to the clerk. The clerk compiled the data from the notes and sent to the central office 'such data . . . as is . . . needed to keep them posted as to the cost and progress of the work and the men's time.'[3]

Two distinctive features of Taylor's system were the bulletin board and written orders or instruction cards, innovations that struck at the foremen's traditional prerogatives of scheduling work and determining the method of production. But Taylor's initiatives, if advanced in comparison to most systematic management plans, were exceedingly crude by his later standards; as he admitted, the instruction card clerks worked under the foreman.

Taylor's approach to supervision was equally uncertain and incomplete. Like other engineers, he was dissatisfied with the foreman's powers and methods. His production control plan attacked their authority indirectly; even in this primitive form the planning office and departmental clerks stripped them of much of their ability to plan and to control production. But Taylor also anticipated a later innovation by assigning some of the foremen's remaining powers to specialized supervisors who would perform only one of their major functions, but perform it carefully and systematically. In the mid-1880s he apparently introduced three 'functional' foremen: the 'gang boss,' the 'disciplinarian,' and the 'inspector.' Yet his gang boss appears to have differed little from the conventional subforeman, and the shop foreman doubled as disciplinarian – hardly an innovation. Only when he introduced the independent inspector 'with orders to go straight to the men rather than to the gang boss,' did he break with the customary approach to supervision.[4]

Taylor's other managerial reforms of the mid-1880s were closely related to his technical work. A reorganized tool room and standardized tools were direct results of the metal-cutting studies. Standardized maintenance procedures for machines and belts and a 'tickler' or maintenance schedule were likewise products of Taylor's efforts to achieve optimum machine performance. His rules for oiling machinery (coupled with the fines he levied for violating them) evoked strong opposition from the workers. [. . .]

Taylor later wrote that he devised stopwatch time study in 1883 and established a time study or 'rate fixing' department – presumably an adjunct to his 'central office' or planning department – in 1884. Copley [Taylor's biographer] concluded, probably correctly, that Taylor's initial goal was 'simply that of improving the statistics long used in the shop in connection with setting piece rates for . . . a new job.'[5] Taylor's contributions were the substitution of a stopwatch for the foreman's conventional timepiece, and, more significantly, the division of the work into basic steps or elements, each of which he timed separately. He then combined his figures to find a time for the entire job. By dividing this figure into the workday, he arrived at an optimum production rate. Sellers [Taylor's employer] soon approved his request to employ a clerk to conduct the time studies. Yet 'it was several years before the full benefits of the system were felt,' since 'the best methods of making and recording time observations' and of 'determining the maximum capacity of each of the machines' and 'of making tables and time tables, were not at first adopted.'[6]

In 1884 Taylor applied his 'differential piece rate' to 'part of the work in the machine shop.'[7] Unlike Halsey [a contemporary management expert], who retained the existing piece rate as his base wage, Taylor set a high bonus rate for workers who completed their assignment in the allotted time and a low penalty rate for all others. Thus the Midvale machinists earned either a high wage or a very low wage, one designed to discourage 'even an inferior man.' On one job, for example, the machinist had received a rate of 50¢ per piece and usually turned out four or five finished pieces per day. After time studies Taylor concluded that the man should produce ten pieces per day. He set a new rate of 35¢ per piece if the machinist finished ten acceptable pieces per day (or a wage of $3.50 rather than $2.00 to $2.50 per day), and 25¢ if he completed nine or fewer pieces (or a maximum of $2.25 per day). Taylor claimed that he established his high or bonus rate by adjusting the piece rate upward until the men agreed to cooperate. In reality he was probably somewhat less gentle. According to a close friend and associate, he fired several machinists who refused to increase their pace.

Between 1884 and 1889 Taylor supposedly extended the differential piece rate to many, perhaps most, of the Midvale workers. Yet his records and published statements mention only the yard laborers and plant maintenance men, and there are several reasons why the records probably suggest the actual thrust of his work. First, as chief engineer he was a staff rather than a line executive and had no direct control over the employees with the possible exception of the machinists. Davenport [the factory superintendent] was certainly amenable to his suggestions, but was probably more receptive to radical plans that applied to the eminently replaceable, powerless laborers than to the valuable machine workers. Second, after 1884 Taylor viewed the differential piece rate as a cheap and easy way to increase output. Yet some types of output were easier and cheaper to increase than others. Because of his personal interest in metal cutting, Taylor fully understood the complexities of rate setting for machine workers, particularly machinists. Laborers presented no such problems and were therefore obvious targets for his efficiency campaign. By 1889 he had applied the differential piece rate to such diverse tasks as unloading pig iron and sand, whitewashing walls, painting, and even changing light bulbs. It seems reasonable to conclude that his wage plan – a plan he customarily described as a spur to machine production and an answer to the restriction of machine output – was in practice primarily an answer to the inefficiencies of workers performing manual, mostly unskilled tasks.

Whatever the exact reason, Taylor's emphasis on the laborers marked the beginning of an ambitious experimental project. Believing that the laborers posed fewer obstacles than the machinists, he began a search for 'what constituted a day's work.' As he later wrote, 'what we hoped ultimately to determine was . . . how many foot-pounds of work a man could do in a day.'[8] He assigned an assistant to review the published literature for possible answers. When the assistant failed to find anything of value, Taylor proceeded with his time studies. Though the work proved difficult and the solution elusive, Taylor remained optimistic. For at least fifteen years he expected to discover an appropriate rule or formula that would enable him to set bonus rates for all such workers. Indeed, it is unlikely that he ever fully reconciled himself to the notion that men could be as complicated as machines.

From Taylor's viewpoint the attractive features of his labor measures were the ease with which they could be implemented and their immediate results. Apart from the determination of metal-cutting times, 'rate-fixing' and the differential piece rate required neither concerted attention nor technical expertise. In 'A Piece Rate System,' Taylor assured his readers that in most plants both activities could be undertaken by one man working part-time. Moreover, the rate fixer 'soon becomes so familiar with the time required to do each kind of elementary work performed by the men, that he can write down the time from memory.'[9] This was in fact what happened at Midvale. As one of Taylor's colleagues reported in 1894:

> [Time study originally] necessitated a considerable clerical force, and they had in the thick of it, about 8 clerks in the estimating department, and some of them were first class men. At the present time their files are full of data on which to base estimates, and the entire work of the estimating office is done by a man . . . It hardly seemed to me that it was safe for them to have but one rate clerk, as his unavoidable absence or sickness would be likely to embarrass them.[10]

Nevertheless the results were impressive. Taylor argued that output had doubled or tripled, that the quality of the work had improved, and, despite initial opposition to time study, that the 'labor problem' had been eliminated.[11] Unfortunately there is no other assessment of the workers' response to time study and the differential piece rate. The most that can be said is that they did not strike or resist in any demonstrative way. Apparently they either welcomed the opportunity to earn higher wages or went elsewhere.

Though the least demanding of his Midvale innovations, time study and the differential piece rate were important steps in the evolution of scientific management and in Taylor's career as a factory manager. By 1890 he recognized that time study could be a potent managerial tool. He was even more confident about the incentive wage. His Midvale experiences indicated that it could be a powerful force in stimulating workers to greater output. In this respect his thinking was similar to that of Halsey and other innovative engineers. But there was also a basic difference. Taylor's conclusions derived wholly from his pragmatic effort to increase production rather than from anxiety over the 'labor problem,' the state of American society, the worker's welfare, or some combination of economic and social concerns.

His single-minded emphasis on production became more apparent in the 1890's when he introduced the differential piece rate in other factories.

Taylor's management innovations, intimately related to his technical achievements, thus made him a pioneer in the management movement. His production control and supervisory methods were as thorough as those of other management reformers, and his wage system was an important refinement of the most advanced practice. But if Taylor profited from his intimate knowledge of the shop, he was also a captive of his Midvale experiences. A master of machine techniques, he had little familiarity with other aspects of management at Midvale. His neglect of purchasing and stores procedures and cost accounting, not to mention corporate finance and marketing, was the most obvious evidence of his limited perspective.

In one other important area Taylor was less advanced than many of his contemporaries. At a time when personnel work was emerging as a distinctive function, Taylor clung to the 'rule of thumb' methods that he condemned in other spheres. He was indifferent to new methods of handling workmen and hostile to welfare work and collective bargaining. This lapse, like most aspects of his work, was a product of his background and professional associations. Sellers was known for his rough treatment of subordinates. Brinley and Davenport were somewhat kinder, but, like Sellers, had little or no interest in personnel management as a specialized activity. Midvale had a mutual benefit insurance society by the late 1870's and a haphazard apprenticeship program, but nothing else. The foremen hired and fired and, except for Taylor's limited use of functional supervisors, controlled their subordinates in the traditional fashion. The management discouraged union membership but did not deliberately exclude or discharge union men. There were no strikes until the 1890s.

Taylor's approach to personnel management was essentially the Midvale approach. He was convinced that men worked for money and that anything that obscured or confused this basic fact was unsound. Forestalling or eliminating the 'labor problem' was an engineering problem, a facet of the larger challenge of systematic production management. Beyond this he had nothing to offer; indeed, in most respects he was a reactionary. In later years he proclaimed the 'scientific selection and progressive development of the workmen' as one of his 'principles' of scientific management, but there were largely meaningless phrases. Taylor's methods of 'selection' and 'development' were to offer bonuses to able and cooperative workers. Those who succeeded 'selected' and 'developed' themselves. His only concession to personnel work was the 'disciplinarian' who, he later explained, might also serve as an employment manager.[12] Welfare work, he believed, was a 'joke.'[13] And under scientific management, labor unions were irrelevant.

[. . .]

Notes

1 Taylor, 'On the Art of Cutting Metals,' ASME *Transactions* 28 (1906): 38.
2 Shartle to Fannon, May 2, 1916, Frederick W. Taylor Papers (Stevens Institute of Technology, Hoboken, N.J.), File 14H.
3 Taylor, 'Discussion,' ASME *Transactions* 7 (1885–86); 475–76.

4 Taylor, 'Shop Management,' ASME *Transactions* 24 (1903): 1396.
5 Frank Barkley Copley, *Frederick W. Taylor, The Father of Scientific Management*, 2 vols. (New York, 1923), 1: 231.
6 Taylor, 'A Piece Rate System,' ASME *Transactions* 16 (1895): 869.
7 Ibid. 876.
8 Taylor, *The Principles of Scientific Management* (New York, 1911), 55.
9 Taylor, 'A Piece Rate System,' 871.
10 Steele to James Deering, April 28, 1894, Taylor Papers, File 70G. There were also serious methodological deficiencies.
11 Taylor, 'A Piece Rate System,' 870.
12 Taylor, 'Shop Management,' 1404.
13 Frank B. Copley, 'Frederick W. Taylor, Revolutionist,' *Outlook* 3 (Sept. 1, 1915): 42.

5.3 Heather Hogan, *Industrial Rationalization and the Roots of Labour Militance in the St Petersburg Metalworking Industry 1901–1914*, 1983

The impact of technological change on the workplace has become an important and fruitful area of investigation. Students of West European and American labour history have begun to chart the complex patterns embedded in the processes set in motion by the 'second' industrial revolution and the managerial reform movement and to link these diverse changes to the upsurge of labour protest on the eve of World War I. While a range of industries have been studied, much of the literature concerning the rationalization of factory operations concentrates specifically on metalworking. Few sectors were influenced as significantly or as early by these diverse trends, and it was in the machine shop and among highly skilled metalworkers that the ideas of Frederick Winslow Taylor took root and matured. Although specifically Taylorist methodologies were rarely employed *in toto* in the United States or elsewhere, the underlying principles of 'scientific management' came to influence major aspects of the work process in this industry; ultimately, the resultant changes in the organization of work played an important role in metalworker struggles from Clydeside to Turin.[1]

While Russian historians have yet to address these issues, the distress caused by industrial rationalization may have significantly informed labour unrest in the interrevolutionary period and lent a special urgency to demands for workers' control in 1917. On the eve of the war, labour protest had become particularly marked in the metalworking industry of St Petersburg and displayed an intensity and tenacity that was rarely matched by other workers in the capital or by workers elsewhere. Leopold Haimson has argued that labour protest among metalworkers was, in part, the product of the dramatic influx of peasants, as well as

Source: *The Russian Review*, vol. 42, 1983, pp. 163–190.

young urban workers who came of age after 1905. Haimson emphasizes the special mix of skilled and unskilled, experienced and inexperienced workers as a factor contributing to this unrest.[2] This paper seeks to build on Haimson's thesis by examining the financial, organizational, and technological changes in the metalworking industry that not only permitted employers to utilize an increasing number of young and untrained workers, but fundamentally altered the terms of labour-management conflict as well. Arguing that the processes of industrial rationalization and employer mobilization effectively undermined the ability of workers to resist these changes at the enterprise level, this investigation interprets the patterns of politicization and strike militance emerging in the pre-war period as the product of a new relationship between capital and labour which directly threatened the interests of the Petersburg *metallisty*.

The first great period of growth for the St Petersburg metalworking industry came in the 1890s and was in large measure the product of Witte's program for rapid industrialization. Stimulated by lucrative orders for rails and rolling stock and protected by high tariffs from foreign competition, heavy industry developed at breakneck speed. The number of metalworking factories rose from 104 in 1890 to 163 in 1900, while the labour force in these factories grew from some 24,600 workers in 1890 to about 64,500 in 1900.

In these tumultuous years, many of the metalworking plants that had been founded decades earlier for the manufacture of diverse products shifted their production to serve the various needs of the railroads. Old plant and auxiliary services were not fundamentally reconstructed; rather, new equipment and new shops sprang up in haphazard fashion in the rush to capitalize on the booming economy. Equipment was bought with an eye to profitable orders with little attention to long-term needs. Plant layout was often chaotic; connecting track between shops tended to be inadequate and poorly laid out, while floor plan was often ignored, and shops were congested with new machinery. The workforce expanded significantly as well, and was perhaps the principal factor in the increased output of these plants, since electrical and mechanical power was still limited. [. . .]

Early descriptions of metalworking factories indicate a strict hierarchy of skills and the subdivision of large factories into numerous, more specialized shops. Nonetheless, metalwork still rested on the skilled use of hand tools and on craftsmanship, and was characterized by a tendency to fabricate whole products. In this period, the metalworker was not a semi-skilled operative, nor did he confront complex and fragmented technological processes. Skill in the machine shops and strength in the metal-processing departments were the stock in trade of the metalworker.

Between 1901 and 1910, two factors coalesced to alter these patterns. First was an industrial depression followed by a period of economic stagnation; second was the outbreak of labour protest in 1901, followed by the far more threatening collapse of labour discipline that attended the workers' revolution of 1905. Industrialists had to reexamine the financial and technical structure of their operations and to rethink the premises of labour-management relations. In response to the strains of this decade, new patterns emerged, reflective of a developing managerial strategy. Elements of this strategy included a marked tendency toward industrial consolidation and employer association, the financial reorganization of many St Petersburg metalworking plants, a reform of business procedures, and a rationalization of work processes. [. . .]

St Petersburg heavy industry was further shaped by the role of foreign technical expertise. Contracts with foreign companies tended to specify the installation of equipment manufactured abroad, and often included provisions for Russian engineers to study at the parent or interested firm. Thus Russians came into contact with the advanced technologies and methods of organization prevalent in the West; they returned home able and perhaps eager to install such equipment and modes of organization in their own factories. These contacts also introduced them to the ideas of the scientific management movement, then enjoying great popularity in America and Western Europe. The technical journals of the 1905–1914 period frequently contain reports of engineers on their tours abroad and discuss various aspects of industrial organization and managerial thinking. Foreign influence was probably strongest in St Petersburg heavy industry and in the new industrial regions of the South and provided a continuing stimulus to technical and organizational modernization.

Equally important in shaping heavy industry in the 1901–1910 period was the changing pattern of state entrepreneurship. Forced to curtail its investment in rapid industrialization, the government could no longer offer lucrative contracts to the now-struggling heavy industry of the Empire. Moreover, as Stolypin replaced Witte, and as agrarian interests gained control of the Duma, government policy began to focus more clearly on the needs of agriculture. Industry began to lose the special encouragement it had enjoyed in the 1890s. Confronted with a sharp reduction in state demand – temporarily, as it would turn out – metalworking industrialists were forced to look to the private, internal market to sell their goods. The turn to this market was modest and halting and could not be otherwise, given the structure of the economy and the weak demand for the products of heavy industry. But this limited demand, and the unfamiliar challenge of competition, forced manufacturers to develop new methods to bring down the high costs of production.

What emerges is a set of conflicting patterns and pressures, which encouraged, on one hand, consolidation and cooperation among industrialists and therefore reduced the need to rationalize factory operations, and, on the other, promoted efforts to cut costs and increase productivity in response to more pronounced competition. The sheltered position of metalworking tended to retard the rationalization of these plants, while the economic dislocation of the first decade of the twentieth century, the emergence of competing industrial conglomerates, and the influence of foreign investment and technical assistance tended to encourage it. The result was a complex commingling of modernity and backwardness in industrial plant. Nonetheless, a more 'modern' industrial structure was beginning to grow out of the old factory regime; Petersburg heavy industry was embarking on an important period of transition. The consolidation of the resources and financial structures of many metalworking firms enhanced the ability of employers to resist the demands of labour. Employers' willingness, indeed need, to resist would be spurred by the shocks of labour militancy in 1905 and again in the 1912–1914 period.

The Revolution of 1905 compounded the crisis of the metalworking industry by undermining the structure of managerial authority in the factories. If the economic crisis of the first decade of the twentieth century called into question the viability of many of the capital's industrial enterprises, the workers' revolution of 1905 provided an equally harsh criticism of the established premises of labour-management relations. And if industrialists responded to the economic pressures of these years by elaborating new forms of industrial consolidation and by embarking on the financial and technical reorganization of their operations, their response to labour protest was in part a reactive and defensive posture, which found expression in an aggresssive anti-labour employer organization and in the gradual articulation of an offensive strategy of industrial rationalization designed to decrease dependence on skilled, often militant metalworkers. [. . .]

As industrialists began to take stock of the damages inflicted by the revolution, they registered important losses. Their concessions on wages and hours intensified the pressure to reduce the wage bill and increase productivity. Further, the newly-won rights of labour to organize in trade unions and strike peacefully seemed to compel management to accept organized labour as a participant in the definition of working conditions. Hence factory owners had not only to restore labour discipline, but also to articulate a new labour-relations policy that might moderate the demands of a more organized workforce.

Also unsettling had been the government's apparent willingness to sacrifice the interests of industry to those of state security. Suddenly the factory owners stood accused of provoking the workers' revolt by the deplorable conditions in the factories. Not only was the autocracy unable to protect industry – to stop the strikes and the assault on managerial authority – but state and society seemed ready to offer up industry as the whipping boy. Thrown on the defensive, industrialists tended more than ever to perceive themselves as an embattled, isolated group. They felt compelled to mobilize their heretofore disparate forces in an employers' organization – the Petersburg Society of Mill and Factory Owners (*Peterburgskoe obshchestvo zavodchikov i fabrikantov*, hereafter PSMFO).[3]

The dual crisis of the metalworking industry provided employers with a motive to rethink established methods of work, not only from a perspective acutely sensitive to economic constraint but from the experience of recent and profoundly troubling social conflict. In part, the resolution of the problems they faced centered on the forging of a united response to labour's demands within the framework of the PSMFO, and, while the significance of this organization was great, attention will focus here on the changes employers began to introduce in the factories between 1906 and 1914.

These reforms embraced four areas. First was a new wage policy; second was an increased concern to influence the ways in which factory time was utilized. Third, various commercial, bookkeeping, and accounting procedures were reorganized and diverse production-line data were collected and systematized. Finally, labour-saving technology was gradually introduced, which simultaneously increased productivity and altered the composition of the workforce. Management was groping toward a strategy that would reduce costs and increase productivity, undermine worker solidarity, instill a new labour discipline, and permit the utilization of a less skilled workforce. As these new approaches to factory organization unfolded, reports in the metalworkers' press documented the distress experienced by workers as their work environment was reshaped. [. . .]

In most St Petersburg metalworking factories, a new wage policy was introduced in the 1906–1910 period and variously consisted of the employment of 'progressive' wage incentive schemes, changes in the unit of time by which wages were calculated, and the introduction of new procedures for recording rates and issuing work. [. . .]

In the labour press, the *'amerikanka'* connoted any premium system built on economies of time and sometimes referred to the use of work cards, order sheets, 'blanks,' or the like, which specified the time in which a task was to be performed. Using this broad definition, at least eleven plants may be identified as employing wage incentive schemes in the 1906–1914 period: Feniks, Kreiton, San-Galli, Vulkan, Struck, Siemens-Halske, Obukhov, Baltic, Nevskii, the Pneumatic Machine Plant, and the Sestroretskii Arms Factory. This evidence is supplemented by a survey conducted by the Metalworkers' Union in 1909, which noted that 'a whole series of attempts were made to transfer to the American system in the 1907–1908 period.'

Throughout the 1906–1914 period, other factories adopted new procedures for rating, issuing, and recording work. By some accounts, the establishment of rates bureaus became the *idée fixe* of many industrialists; in the capital bureaus existed at the Lessner, Nevskii, Nobel', New Admiralty, and Semenov plants, and perhaps at Obukhov and Baltic. Workers were threatened and concerned by such bodies. The pace of work was not only substantially intensified, but the ability of workers to influence rate determination was diminished. The rates bureau depersonalized the rate-setting process by intervening between the worker and his foreman. The worker could no longer bargain with the foreman directly, but was told to direct his complaints elsewhere or informed that he had no reason to complain since rates had been set 'scientifically.' Nevskii workers pointedly expressed their feelings in regard to such bodies by calling theirs the 'funeral bureau,' a derisive name that reflected not only scorn but a sense of hopelessness.

Further innovation was apparent in a new array of printed forms given to the worker when he was issued a task. These were variously called order sheets, control sheets, or a check or card system. Typically, the worker had to notc when he began and ended a task. Some forms specified the time allotted for a task, others defined the procedure and materials the worker was to use. These new requirements also generated anxiety within the workers' ranks. Thus in 1914, workers at the Putilov Wharf downed tools when an order system and notation by tags was instituted in place of the old method of calculating piece rates according to receipts. This, complained one worker, allowed the administration to calculate costs on work individually and to lower rates that it deemed too high, thereby increasing the intensity of labour.

Most St Petersburg plants replaced the day wage with an hourly wage following 1905, and many employed wage incentive schemes rather than simple piece rates. Innovative payment plans often entailed the institution of special rates bureaus and the utilization of new printed forms. These reforms were frequently part of a larger process of reorganization, including the expanded employment of technical and office personnel to both record and document the work process and to increase supervision over the labour force. A 'rational' wage policy promised management reduced costs and increased productivity. Workers, though, perceived these changes as falling piece rates, a faster pace of work, and a new and disconcerting discipline.

As management began to scrutinize more closely the problems of cost and productivity, the question of factory time came into sharper focus. Attention to the systematic 'soldiering' of production-line workers led employers to examine the pace and structure of the work day more generally. Major reforms in the area of time management included the abolition or restriction of the traditional 'free time' – that five to fifteen minutes before the start of work in the morning or after dinner when workers conversed, put their tools in order, or simply relaxed without fear of a fine. Automatic timeclocks began to appear in many of the capital's metalworking factories, which 'red-lined' minutes of lateness and afforded a strange contrast to the old practice of hanging 'numbers' – square or oval tags – on hooks on a board. Workers complained bitterly that such methods lengthened the work day: Putilovtsy [Putilov workers] destroyed a newly installed 'fine' bell, while angry Obukhov workers hurled rocks at the turnstiles that led to the new timeclocks. Equally troubling was the intensified supervision of foremen:

> The strengthened staff of foremen vigilantly follow each move of the worker: God save you if you talk with a neighbour, take a smoke, leave the bench, sit in a group, or still worse, glance at a paper. Fine, Fine, Fine!

Some workers grumbled that this supervision even extended into the bathroom, while others felt that the new restrictions on tea and *kipiatok* [boiling water] in the shops was detrimental to the workers' health. And the practice known in early American textile mills as 'the stretch-out' – work on more than one machine or operation at the same time – was now employed by some metalworking plants of the capital.

Perhaps most disturbing was the arrival of 'time-work' specialists on the shop floor. While it is difficult to determine the extent to which time study was employed in St Petersburg metalworking factories, there is no doubt it was the cause of considerable anxiety within labour's ranks and that the personnel who practised it were often met with anger and resentment. At the Glebov Factory the director had his 'foremen go around with watches and check up on the workers,' while at the Metal Factory a worker complained that 'our fattened foreman comes up to the bench, takes out a watch, and says to the workers: "set it with more traction, at a faster pace and take it with greater force."'

Young technical students on assignment in the factories were often the practitioners of time study; their age and their methods combined to give particular offence to the workforce. At the Baltic Factory such students, acting as assistant foremen, administered tests to incoming workers and observed their performance with stopwatches. The Nevskii Factory employed students in its rates bureau, and it was these young technicians who studied the machine tools and the workers, and who then calculated speeds and wage rates. In 1913, a sixty-day strike at the New

Aivaz Factory broke out over the issue of time study, a conflict that began when the student Balik was carted out of the shop in a wheelbarrow.

Taken together, management's heightened concern to influence the utilization of factory time deeply affected the patterns that had once defined the work day. The customs that had animated the shops and grew in the moments of a break were gradually circumscribed, to be reshaped by the more exacting requirements of the punchclock and stopwatch. Management sought structures to instil a new discipline, methods to break peasant-workers into the rhythms of industry, ways to compel skilled workers to stay at their benches. Employers believed they were combatting entrenched attitudes and customs that impeded the development of industrial Russia. But for workers, the intrusion of management ever more deeply into the shop culture was both threatening and demoralizing, and many would struggle to thwart this incursion.

Integral to many reforms in wage policy and time management was the collection and analysis of production-line data. However, the business practices of an earlier era were inadequate to these new tasks. Few managers had records on hand to document expenditures on materials and labour; few knew in any detail the costs of production; few had bothered to monitor work flows carefully; few employed the requisite technical and administrative staff to generate this information. The specialized press provided practical guidelines for the implementation of the most basic reforms in business practice and bemoaned the primitiveness of many accounting procedures. It also offered a plethora of forms to serve as models for the reorganization of business practices. Often the substitution of written for oral communication stood at the centre of such reforms, designed to systematize business operations, provide simple and clear indices of productivity, and develop a greater responsiveness at all levels to managerial directives. Diverse aspects of the commercial and production process could now be more closely scrutinized. Formerly, accurate records on the cost of producing individual articles were rarely maintained, but now management was introduced to cost-benefit analysis and more refined accounting methods. Earlier, pay office clerks laboured over torn and dirty workbooks and the barely legible notations of foremen; now worksheets supplemented the workbook, and this far more detailed printed form gradually allowed for the gathering of crucial data on the organization of the production process and permitted it to be passed 'upward' to management for analysis and subsequent action. New forms specified the worker's name and number, the rates and time of work, the materials and instruments used, and sometimes the methods by which work was to be performed. Further systemization included the installation of timeclocks, with the punchcard constituting an additional record of account with the worker. Office procedures became more specialized, and simple mechanical devices were applied to routine operations. Filing systems organized blueprints and drawings for the needs of the planning department; heretofore such documents had remained in the shops, under the physical control of foremen and workers.

The collection and cataloguing of such material necessarily entailed the expansion of technical and administrative staff. Although adequate statistical data on the employment of such personnel are generally lacking, there is evidence of their increasing role in St Petersburg metalworking factories. Perhaps most telling from the point of view of labour-management conflict were the numerous reports in the labour press documenting the increasing tension that accompanied the expansion of supervisory personnel throughout the 1907–1914 period. Such an increase was especially felt between 1907 and 1910 when the workforce was cut due to the curtailment of production, while the administrative staff remained the same or was enlarged. This trend was further evidenced by the promotion of workers to the post of foreman. Workers frequently complained that 'former comrades,' now working for management, had betrayed them; their hostility found pointed expression in a 1910 decision to exclude such workers from the Metal Workers' Union. Finally, the expansion of technical education made an increasing number of trained professionals available to industry; between 1896 and 1902 the number of higher technical schools doubled, as did enrolment at existing institutes, producing in all a four-fold increase in the number of engineering students.

Functional specialization resolved another problem in factory administration: the training of foremen and managers. Stressed one employer, 'Up to now the typical foreman was a universal man who acquired those skills that permitted him to direct work by long years of experience. His knowledge became an art and since his sphere of competence was great, then, generally speaking, the productivity and quality of his work was not great.'

The Semenov Factory – widely considered a model machine-construction plant – provides one of the best illustrations of the integration of diverse aspects of the rationalization process: the utilization of advanced technologies, the subdivision and functional specialization of work and staff, the increase in written documentation to structure factory operations, the expansion of administrative staff with an attendant increase in overhead expenses, and the application of elements of Taylorism (e.g., time study, progressive wage systems). The goal of these reforms was the elimination of a factory's 'accidental dependence' on the human personality and the substitution of a 'system' for the discretion of individuals. Managerial control replaced the diffuse control exercised by the various participants in the work process. The 'art' of the skilled foreman and worker and the value of long years of training and experience were degraded.

The application of advanced technologies to the work process was a constituent element in the rationalization of the metalworking industry. As production processes were mechanized and standardized, a more refined division of labour became possible. The subdivision of work led to the simplification of individual operations, which in turn permitted employers to hire less skilled labour. The role of the skilled worker in the factory was narrowed. Caught between an expanded cadre of semi- and unskilled labour 'below' him and an enlarged contingent of technical personnel 'above' him, the skilled metalworker confronted a severe loss of status and sharply reduced possibilities for advancement. For management, such a developmental strategy promised to relieve the industry of its dependence on the skilled, highly paid and often militant metalworker. In his place, factory owners could tap Russia's abundant reserves of unskilled, female, and adolescent labour – labour that was cheaper and presumably more docile. To be sure, innovation in these areas was gradual, indeed only just beginning, and the impact on workforce composition was still relatively modest. Nonetheless, workers feared the general implementation of a restructured workforce.

Reports from the factories in the immediate post-1905 period began to register workers' alarm, as management turned to unskilled cadres. Women replaced men at the Cable Factory, and adolescents took over adults' jobs at the Cartridge-Case Plant. In some places common labourers were put on machine tools, not to train them to become metalworkers but 'with a view to economy.' Other plants expanded the use of adolescents, and when adult workers left or were fired, they were replaced by assistants 'who, after a year and a half of work, were already considered genuine lathe operators.' At the Pipe Factory automatic machine tools were installed, which not only lowered rates but, workers assumed, would soon displace skilled workers. The use of common labour in place of skilled was noted as well in the tram repair parks and at the Vulkan Factory. In yet another case, workers at Siemens-Halske complained that the administration, trying to reduce costs as much as possible, had placed inexperienced workers 'who perhaps have never seen a machine tool before' on the benches and this had 'naturally' led to the increase in accidents at the factory.

Throughout the interrevolutionary period, the labour press voiced concern about these emerging trends. A July, 1907, article spoke of 'major technological changes that are displacing workers,' while an informal survey of thirty-eight plants conducted by Malinovskii in the late summer of 1909 sketched the following picture. These plants had reduced their workforce by seven percent in the past year and a half. Virtually all dismissals had been accompanied by a decrease in rates, and at the same time adult workers were replaced by adolescents, men by women, trained workers by common labourers. While technical modernization was sporadic, various improvements were noted in the forge shop at Putilov (new riveting presses had dramatically increased the speed of this operation), new machine tools and cranes were introduced at the Cable Factory and at Langenzippen, and a range of improvements had occurred at the Aleksandrovskii Plant.

Unfortunately, available statistical data on female employment are not particularly illuminating. But two sets of figures offer a crude index. *Metallist*, citing factory inspectorate material, reported that the number of women in metalworking had increased by 33 percent between 1901 and 1910, while the male component grew by only 8 percent. Similar data for 1913, exclusive of the state plants, indicated a 3.3 percent rate of female employment, up from the 1.3 percent reported in the more comprehensive city census of 1900. Articles in *Pravda* from late 1913 and early 1914 suggest that 10,000 to 15,000 women were employed in metalworking, thereby constituting 8–12 percent of the workforce.

While female employment remained modest until the war, many metalworkers were more troubled than the actual number of female metalworkers might warrant. This anxiety rested, in part, on the 'intrusion' of women into a once exclusively male world; also at issue was the perceived threat of displacement and dilution. Thus for some, the factory began to seem like an 'odious women's city,' and many 'grey' workers were quick to curse the 'old women' who 'get in' everywhere. Others evinced a greater sensitivity to the plight of women in the shops. Commented one metalworker: 'To our shame, the relations of the rank-and-file worker to women are frequently shockingly crude.' These concerns were manifested in other ways as well. In the pre-war period, the labour press carried an increasing number of articles on female labour, while the Metal Workers' Union made an effort to recruit members among women by lowering dues. Another striking indicator of shifting attitudes was the election of two women to the governing board of the union in 1913. And, probably as much a reflection of the interests of skilled workers undergoing a process of displacement as a sensitivity to the needs of women, strike demands began to insist on minimum wage rates for female metalworkers.

Similar concerns came to light in regard to apprenticeship programs. [. . .] As work processes were simplified, training could be accomplished in a much shorter period of time; and rather than serving to teach a worker the skills vital to his trade, management now held workers in the apprentice category longer than necessary to secure cheap labour. As young workers entered the shops in increasing numbers, tensions sometimes flared; not a few apprentices suffered beatings at the hands of adult workers.

Also indicative of the simplification of work processes and the expanded utilization of unskilled cadres was the lowering of piece rates. [. . .] Thus in the most technically advanced metalworking factories – leading machine construction plants like Nobel', Erickson, New Aivaz, and Vulkan – protests against rate cuts of as much as 30–50 percent occurred in the prewar years. At a meeting of the board of the Metalworkers' Union in November, 1913, moreover, workers reported that the majority of recent strikes had been caused by decreases in rates.

Diverse processes thus combined to alter the workforce and to erode the male camaraderie and pride in skill that had heretofore defined the shop culture of Petersburg's metalworking factories. The long-established customs and rhythms of the work day seemed now to be changing quickly, as the plants of the capital traversed the distance between the old factory regime and the 'modern' industrial enterprise. These processes were scarcely complete by 1914; however, they were still unfolding. Yet resistance to these trends was beginning to assume a central place in the strike movement of the Petersburg metalworkers.

Many factors contributed to the pre-war militance of the *metallisty*, not least of which was an erosion of the established patterns of deference and obedience, reflected so well in worker demands for polite forms of address. Also important were the Duma forum, the labour press, and various labour organizations, all of which expanded the avenues open to workers for the articulation of their common grievances. Significant too, was the frustration experienced by many workers when their deputies were silenced, their leaders harassed by arrest, exile, or imprisonment, and their organizations closed by an all-too-watchful autocracy. In these years, moreover, the economically and politically motivated strikes of metalworkers became closely intertwined, and it seemed that the workers' search for economic and social rights was leading to a confrontation with an archaic political regime. But in important respects, this assault on the established premises of state and society tended to obscure the conflict surrounding the emergence of a substantially new socio-economic order. The rationalization of work processes and the consolidation of the financial, commercial, and industrial structure of the St Petersburg metalworking industry was symptomatic of the larger capitalist transformation of Russian society. The fierce, and by 1914 increasingly political, protest of the *metallisty* was motivated in large measure by efforts to check or slow this transformation. Thus to the volatile mix of factors that agitated metalworkers on the eve of war and revolution, an intense conflict over the shifting relations of labour and capital must be added.

One must stress again that the processes of industrial rationalization and employer mobilization were only just beginning to unfold and that worker resistance to these trends was typically indirect and rarely expressly articulated. Protest against the lowering of piece rates, various demands in regard to women, apprentices, and the unskilled, as well as worker insistence on paid leaves and a reduction in work hours were, however, indicative of the struggle against the substantial reorganization of work relations that had occurred since the first revolution.

More explicit was the resistance of New Aivaz workers. In the summer of 1913, the accumulated tensions generated by management's programme of technical rationalization, time study, rate cuts, the expanded employment of women and unskilled labour, and the displacement of skilled metalworkers erupted in a sixty-day strike. Appealing to Petersburg workers for support, the Aivazovtsy were anxious to alert their comrades to the changes that would soon affect all *metallisty*:

> New Aivaz is the first major factory where the American system of Taylor has been introduced, with its division of labour, machine-ism, and time study. Our conflict is a struggle against this approaching new refinement of slavery, against a callous, merciless oppression. To win our strike is to secure all Petersburg from the savage extremes of the American system, to take the first step in the subsequent difficult and crucial struggle. What will become of New Aivaz if the workers of Petersburg leave it without their help? It will not be a factory, but a 'workhouse' with its horrible bondage; it will be a brutal laboratory of human exploitation, where the latest work in technical science will be bought at the price of hunger, humiliation, sweating, illness, and premature death.

The Aivazovtsy would lose their struggle and, in the months following the defeat of their strike, work processes would continue to be transformed at a rapid pace. But the issues raised by this conflict would subsequently be taken up by labour activists on the pages of the union's journal, *Metallist*. A two-part article by Shliapnikov in the fall of 1913 castigated Taylorist methods as a system less concerned with technological progress than with advanced forms of capitalist exploitation. An article signed 'A. Zorin' probed more deeply and sought to define the various elements that constituted a 'factory revolution.'

> Frequently, a strong argument is raised against female metalworkers: women, it is said, will reduce wages, women arrived at the vices just when men insisted on good piece rates. Yes, all this is true. But new strata of workers are called to the factories only because this is advantageous to capital. The same was observed at factories when the unskilled worker ousted the trained worker from almost all positions.

> The arrival of women at the vices, therefore, was inevitable. If not today then tomorrow, if not in 1913 then 1914. Machine tools are modernized, the division of labour proceeds all the more deeply and broadly, work is increasingly simplified, and consequently, the capitalists' appetite for cheap, untrained labour, among whom are women, grows with every passing day. Female metalworkers, therefore, are the inevitable result of all technological development, of the entire capitalist system. It would be improvident, short-sighted to fight women and drive them from the vices. It is necessary to intensify the struggle against the entire system, it is necessary to organize ourselves more strongly for new militant efforts. Even the blind know that our bourgeois world is governed by only one law: the pursuit of profits, the pursuit of gain.

'Zorin's' article contained a near classic description of the impact of technological change. He spoke variously of the destruction of 'metal-working standards,' the denigration of apprenticeship, the passing of the generalized and multi-competent worker and the creation of a specialized labourer whose working gestures were simple, monotonous, and unthinking. Perhaps most importantly, he urged his comrades not to curse women or to heckle the unskilled as '*skoptsy*,' [eunuchs] but to comprehend the underlying reasons for these changes.[4]

It was precisely here that 'Zorin' articulated the conflict enveloping the metalworking industry. Industrial rationalization, he argued, was a necessary and inevitable product of capitalist development. The distress this caused would intensify and consume ever larger segments of the working class. The only way metalworkers could lessen the dislocation brought about by these processes would be by challenging the socio-economic system that benefited from them. 'If not today then tomorrow' all metalworkers would be affected by the changes that had occurred at New Aivaz and thus must mobilize broadly and politically in response.

These were the perceptions that animated a growing number of metalworkers on the eve of World War I and that fuelled the intense strikes in this industry. Since the turn of the century, the structure of the metal-working industry had changed significantly and as a product of industrial rationalization, monopolization, the expanding role of a small number of banking institutions, and the growing influence of the Petersburg Society of Mill and Factory Owners, employers had been able to resist the demands of labour. By 1914, metalworkers were suffering a remarkable rate of strike failure at the enterprise level, and frustrated by defeat after defeat on the shop floor, their protest became increasingly political. Metalworkers rose to protest not only the specific changes in the factories or the repressions of a decaying autocracy, but struggled as well against the emerging power of an important segment of the industrial bourgeoisie. The acute tension in St Petersburg on the eve of war and revolution derived from a conflict over the nature and future relationship of labour and capital.

Notes

1 The literature is substantial and growing, but especially important studies include D. Montgomery, *Workers' Control in America*, Cambridge, 1979; James

Hinton, *The First Shop Stewards' Movement*, London, 1973; and Dieter Groh, 'The Intensification of Work and Industrial Conflict in Germany, 1896–1914,' *Politics and Society*, vol. 8, no. 3/4, 1978, pp. 349–397.

2 Leopold Haimson, 'The Problem of Social Stability in Urban Russia, 1905–1917. Part One,' *Slavic Review*, vol. 23 (1964), esp. pp. 633–637.

3 The history of this organization is discussed in detail in Heather Hogan, 'Labor and Management in Conflict: The St Petersburg Metalworking Industry, 1900–1914,' Ph. D. Diss., University of Michigan, 1981, esp. Chapters 5 and 10.

4 Ironically, 'A. Zorin' was one of the pseudonyms of Aleksei Gastev, later the head of the Central Labour Institute and a popularizer of scientific management in the early Soviet period. See, Kendall E. Bailes, 'Alexei Gastev and the Soviet Controversy Over Taylorism, 1918–1924,' *Soviet Studies,* vol. 29 (1977), pp. 373–394.

6
Science, Technology and Society

6.0 Introduction

The empirical studies in the previous chapters have turned up much to disconcert anyone who maintains that society is shaped by an autonomously-evolving, science-based technology. Surely not all social change is due to technology, and surely technology itself has its social roots – these are objections any technological determinist must accommodate.

Robert L. Heilbroner effectively concedes the first point, but deflects much of the force of the second. The detailed social contextualization of technological innovation and diffusion provided in some of the previous readings would on his account simply amount to a more precise specification of technology's determining roles. Social influences on the applications of technology in any case take second place to an apparently autonomously-evolving general pattern of technological development. Heilbroner's sympathy with technological determinism is deceptive; in the end, the potency of technology is shown to be itself a product of a particular socioeconomic order, and we may look forward to a time when this power will be curbed.

Heilbroner's identification of Marx as a technological determinist now looks problematic, but views more pessimistic than his own about the power of modern science-based technology have emerged from revisions of the Marxian legacy in Western Europe. The view that the technology of modern industrial societies has advanced to a point where it can neutralize political opposition was argued influentially by Herbert Marcuse in *One-Dimensional Man* (1964). This line of thought has been developed by Jürgen Habermas, at one time like Marcuse a member of the Frankfurt Institute for Social Research. His theme in the edited article which ends the volume is the submersion of 'traditional society' by capitalism's ever-expanding 'subsystems of purposive-rational action'. A static society, in which the existing order was underpinned by a set of religiously-based values, has given way to a dynamic one, in which social problems are allowed only 'technical' solutions. It is a society in which science and technology have become the leading productive forces; this is a state of affairs legitimated by a 'technocratic consciousness' which conceals the social interests behind technological progress, and perpetuates the myth that the socioeconomic order is the inevitable outcome of an autonomously-developing science and technology.

Beneath Habermas's difficult prose we may detect a conclusion surprisingly similar to Heilbroner's, despite their radically differing conceptions of science and technology. Neither subscribes to technological determinism in an 'absolute' sense: society cannot *always* be read off the prevailing technology. The plausibility of technological determinism is historically contingent, and stems largely from a rapid growth in the power of technology following its harnessing to science during the 'second industrial revolution'.

If either author embraces technological determinism, it is a qualified, 'progressive' variant: one which, whatever its theoretical shortcomings, is at least not quite as open to simple, empirical refutation as its predecessors.

6.1 Robert L. Heilbroner, *Do Machines Make History?* 1967

> The hand-mill gives you society with the feudal lord; the steam-mill, society with the industrial capitalist.
> Marx, *The Poverty of Philosophy*

That machines make history in some sense – that the level of technology has a direct bearing on the human drama – is of course obvious. That they do not make all of history, however that word be defined, is equally clear. The challenge, then, is to see if one can say something systematic about the matter, to see whether one can order the problem so that it becomes intellectually manageable.

To do so calls at the very beginning for a careful specification of our task. There are a number of important ways in which machines make history that will not concern us here. For example, one can study the impact of technology on the *political* course of history, evidenced most strikingly by the central role played by the technology of war. Or one can study the effect of machines on the *social* attitudes that underlie historical evolution: one thinks of the effect of radio or television on political behaviour. Or one can study technology as one of the factors shaping the changeful content of life from one epoch to another: when we speak of 'life' in the Middle Ages or today we define an existence much of whose texture and substance is intimately connected with the prevailing technological order.

None of these problems will form the focus of this essay. Instead, I propose to examine the impact of technology on history in another area – an area defined by the famous quotation from Marx that stands beneath our title. The question we are interested in, then, concerns the effect of technology in determining the nature of the *socioeconomic order*. In its simplest terms the question is: did medieval technology bring about feudalism? Is industrial technology the necessary and sufficient condition for capitalism? Or, by extension, will the technology of the computer and the atom constitute the ineluctable cause of a new social order?

Even in this restricted sense, our inquiry promises to be broad and sprawling. Hence, I shall not try to attack it head-on, but to examine it in two stages:

1 If we make the assumption that the hand-mill does 'give' us feudalism and the steam-mill capitalism, this places technological change in the position of a prime mover of social history. Can we then explain the 'laws of motion' of technology itself? Or to put the question less grandly, can we explain why technology evolves in the sequence it does?

Source: *Technology and Culture*, vol. 8, 1967, pp. 335–345.

2 Again, taking the Marxian paradigm at face value, exactly what do we mean when we assert that the hand-mill 'gives us' society with the feudal lord? Precisely how does the mode of production affect the superstructure of social relationships?

These questions will enable us to test the empirical content – or at least to see if there is an empirical content – in the idea of technological determinism. I do not think it will come as a surprise if I announce now that we will find *some* content, and a great deal of missing evidence, in our investigation. What will remain then will be to see if we can place the salvageable elements of the theory in historical perspective – to see, in a word, if we can explain technological determinism historically as well as explain history by technological determinism.

We begin with a very difficult question hardly rendered easier by the fact that there exist, to the best of my knowledge, no empirical studies on which to base our speculations. It is the question of whether there is a fixed sequence to technological development and therefore a necessitous path over which technologically developing societies must travel.

I believe there is such a sequence – that the steam-mill follows the hand-mill not by chance but because it is the next 'stage' in a technical conquest of nature that follows one and only one grand avenue of advance. To put it differently, I believe that it is impossible to proceed to the age of the steam-mill until one has passed through the age of the hand-mill, and that in turn one cannot move to the age of the hydro-electric plant before one has mastered the steam-mill, nor to the nuclear power age until one has lived through that of electricity.

Before I attempt to justify so sweeping an assertion, let me make a few reservations. To begin with, I am fully conscious that not all societies are interested in developing a technology of production or in channeling to it the same quota of social energy. I am very much aware of the different pressures that different societies exert on the direction in which technology unfolds. Lastly, I am not unmindful of the difference between the discovery of a given machine and its application as a technology – for example, the invention of a steam engine (the aeolipile) by Hero of Alexandria long before its incorporation into a steam-mill. All these problems, to which we will return in our last section, refer however to the way in which technology makes its peace with the social, political, and economic institutions of the society in which it appears. They do not directly affect the contention that there exists a determinate sequence of productive technology for those societies that are interested in originating and applying such a technology.

What evidence do we have for such a view? I would put forward three suggestive pieces of evidence:

1 The simultaneity of invention

The phenomenon of simultaneous discovery is well known.[1] From our view, it argues that the process of discovery takes place along a well-defined frontier of knowledge rather than in grab-bag fashion. Admittedly, the concept of 'simultaneity' is impressionistic,[2] but the related phenomenon of technological 'clustering' again suggests that technical evolution follows a sequential and determinate rather than random course.[3]

2 **The absence of technological leaps**

All inventions and innovations, by definition, represent an advance of the art beyond existing base lines. Yet, most advances, particularly in retrospect, appear essentially incremental, evolutionary. If nature makes no sudden leaps, neither, it would appear, does technology. To make my point by exaggeration, we do not find experiments in electricity in the year 1500, or attempts to extract power from the atom in the year 1700. On the whole, the development of the technology of production presents a fairly smooth and continuous profile rather than one of jagged peaks and discontinuities.

3 **The predictability of technology**

There is a long history of technological prediction, some of it ludicrous and some not.[4] What is interesting is that the development of technical progress has always seemed *intrinsically* predictable. This does not mean that we can lay down future timetables of technical discovery, nor does it rule out the possibility of surprises. Yet I venture to state that many scientists would be willing to make *general* predictions as to the nature of technological capability twenty-five or even fifty years ahead. This too suggests that technology follows a developmental sequence rather than arriving in a more chancy fashion.

I am aware, needless to say, that these bits of evidence do not constitute anything like a 'proof' of my hypothesis. At best they establish the grounds on which a prima facie case of plausibility may be rested. But I should like now to strengthen these grounds by suggesting two deeper-seated reasons why technology *should* display a 'structured' history.

The first of these is that a major constraint always operates on the technological capacity of an age, the constraint of its accumulated stock of available knowledge. The application of this knowledge may lag behind its reach; the technology of the hand-mill, for example, was by no means at the frontier of medieval technical knowledge, but technical realization can hardly precede what men generally know (although experiment may incrementally advance both technology and knowledge concurrently). Particularly from the mid-nineteenth century to the present do we sense the loosening constraints on technology stemming from successively yielding barriers of scientific knowledge – loosening constraints that result in the successive arrival of the electrical, chemical, aeronautical, electronic, nuclear, and space stages of technology.[5]

The gradual expansion of knowledge is not, however, the only order-bestowing constraint on the development of technology. A second controlling factor is the material competence of the age, its level of technical expertise. To make a steam engine, for example, requires not only some knowledge of the elastic properties of steam but the ability to cast iron cylinders of considerable dimensions with tolerable accuracy. It is one thing to produce a single steam-machine as an expensive toy, such as the machine depicted by Hero, and another to produce a machine that will produce power economically and effectively. The difficulties experienced by Watt and Boulton in achieving a fit of piston to cylinder illustrate the problems of creating a technology, in contrast with a single machine.

Yet until a metal-working technology was established – indeed, until an embryonic machine-tool industry had taken root – an industrial technology was impossible to create. Furthermore, the competence required to create such a technology does not reside alone in the ability or inability to make a particular machine (one thinks of Babbage's ill-fated calculator as an example of a machine born too soon), but in the ability of many industries to change their products or processes to 'fit' a change in one key product or process.

This necessary requirement of technological congruence[6] gives us an additional cause of sequencing. For the ability of many industries to co-operate in producing the equipment needed for a 'higher' stage of technology depends not alone on knowledge or sheer skill but on the division of labor and the specialization of industry. And this in turn hinges to a considerable degree on the sheer size of the stock of capital itself. Thus the slow and painful accumulation of capital, from which springs the gradual diversification of industrial function, becomes an independent regulator of the reach of technical capability.

In making this general case for a determinate pattern of technological evolution – at least insofar as that technology is concerned with production – I do not want to claim too much. I am well aware that reasoning about technical sequences is easily faulted as *post hoc ergo propter hoc*. Hence, let me leave this phase of my inquiry by suggesting no more than that the idea of a roughly ordered progression of productive technology seems logical enough to warrant further empirical investigation. To put it as concretely as possible, I do not think it is just by happenstance that the steam-mill follows, and does not precede, the hand-mill, nor is it mere fantasy in our own day when we speak of the coming of the automatic factory. In the future as in the past, the development of the technology of production seems bounded by the constraints of knowledge and capability and thus, in principle at least, open to prediction as a determinable force of the historic process.

The second proposition to be investigated is no less difficult than the first. It relates, we will recall, to the explicit statement that a given technology imposes certain social and political characteristics upon the society in which it is found. Is it true that, as Marx wrote in *The German Ideology*, 'A certain mode of production, or industrial stage, is always combined with a certain mode of cooperation, or social stage,'[7] or as he put it in the sentence immediately preceding our hand-mill, steam-mill paradigm, 'In acquiring new productive forces men change their mode of production, and in changing their mode of production they change their way of living – they change all their social relations'?

As before, we must set aside for the moment certain 'cultural' aspects of the question. But if we restrict ourselves to the functional relationships directly connected with the process of production itself, I think we can indeed state that the technology of a society imposes a determinate pattern of social relations on that society.

We can, as a matter of fact, distinguish at least two such modes of influence:

1 **The composition of the labor force**
In order to function, a given technology must be attended by a labor force of a particular kind. Thus, the hand-mill (if we may take this as referring to late medieval technology in general) required a work force composed of skilled or semiskilled craftsmen, who were free to practice their occupations at home or in a small atelier, at times and seasons that varied considerably. By way of contrast, the steam-mill – that is, the technology of the nineteenth century – required a work force composed of semiskilled or unskilled operatives who could work only at the factory site and only at the strict time schedule enforced by turning the machinery on or off. Again, the technology of the electronic age has steadily required a higher proportion of skilled attendants; and the coming technology of automation will still further change the needed mix of skills and the locale of work, and may as well drastically lessen the requirements of labor time itself.

2 **The hierarchical organization of work**
Different technological apparatuses not only require different labor forces but different orders of supervision and co-ordination. The internal organization of the eighteenth-century handicraft unit, with its typical man-master relationship, presents a social configuration of a wholly different kind from that of the nineteenth-century factory with its men-manager confrontation, and this in turn differs from the internal social structure of the continuous-flow, semi-automated plant of the present. As the intricacy of the production process increases, a much more complex system of internal controls is required to maintain the system in working order.

Does this add up to the proposition that the steam-mill gives us society with the industrial capitalist? Certainly the class characteristics of a particular society are strongly implied in its functional organization. Yet it would seem wise to be very cautious before relating political effects exclusively to functional economic causes. The Soviet Union, for example, proclaims itself to be a socialist society although its technical base resembles that of old-fashioned capitalism. Had Marx written that the steam-mill gives you society with the industrial *manager*, he would have been closer to the truth.

What is less easy to decide is the degree to which the technological infrastructure is responsible for some of the sociological features of society. Is anomie, for instance, a disease of capitalism or of all industrial societies? Is the organization man a creature of monopoly capital or of all bureaucratic industry wherever found? These questions tempt us to look into the problem of the impact of technology on the existential quality of life, an area we have ruled out of bounds for this paper. Suffice it to say that superficial evidence seems to imply that the similar technologies of Russia and America are indeed giving rise to similar social phenomena of this sort.

As with the first portion of our inquiry, it seems advisable to end this section on a note of caution. There is a danger, in discussing the structure of the labour force or the nature of intrafirm organization, of assigning the sole causal efficacy to the visible presence of machinery and of overlooking the invisible influence of other factors at work. Gilfillan, for instance, writes, 'engineers have committed such blunders as saying the typewriter brought women to work in offices, and with the typesetting machine made possible the great modern newspaper, forgetting that in Japan there are women office workers and great modern newspapers getting practically no help from typewriters and typesetting machines.'[8] In addition, even where technology seems unquestionably to play the critical role, an independent 'social' element unavoidably enters the scene in the *design* of technology, which must take into account such facts as the level of education of the work force or its relative price. In this way the machine will reflect, as much as mould, the social relationships of work.

These caveats urge us to practice what William James called a 'soft determinism' with regard to the influence of the machine on social relations. Nevertheless, I would say that our cautions qualify rather than invalidate the thesis that the prevailing level of technology imposes itself powerfully on the structural organization of the productive side of society. A foreknowledge of the shape of the technical core of society fifty years hence may not allow us to describe the political attributes of that society, and may perhaps only hint at its sociological character, but assuredly it presents us with a profile of requirements, both in labor skills and in supervisory needs, that differ considerably from those of today. We cannot say whether the society of the computer will give us the latter-day capitalist or the commissar, but it seems beyond question that it will give us the technician and the bureaucrat.

Frequently, during our efforts thus far to demonstrate what is valid and useful in the concept of technological determinism, we have been forced to defer certain aspects of the problem until later. It is time now to turn up the rug and to examine what has been swept under it. Let us try to systematize our qualifications and objections to the basic Marxian paradigm:

1 Technological progress is itself a social activity

A theory of technological determinism must contend with the fact that the very activity of invention and innovation is an attribute of some societies and not of others. The Kalahari bushmen or the tribesmen of New Guinea, for instance, have persisted in a neolithic technology to the present day; the Arabs reached a high degree of technical proficiency in the past and have since suffered a decline; the classical Chinese developed technical expertise in some fields while unaccountably neglecting it in the area of production. What factors serve to encourage or discourage this technical thrust is a problem about which we know extremely little at the present moment.[9]

2 The course of technological advance is responsive to social direction

Whether technology advances in the area of war, the arts, agriculture, or industry depends in part on the rewards, inducements, and incentives offered by society. In this way the direction of technological advance is partially the result of social policy. For example, the system of interchangeable parts, first introduced into France and then independently into England failed to take root in either country for lack of government interest or market stimulus. Its success in America is attributable mainly to government support and to its appeal in a society without guild traditions and with high labor costs.[10] The general *level* of technology may follow an independently determined sequential path, but its areas of application certainly reflect social influences.

3 Technological change must be compatible with existing social conditions

An advance in technology not only must be congruent with the surrounding technology but must also be compatible with the existing economic and other institutions of society. For example, labor-saving machinery will not find ready acceptance in a society where labor is abundant and cheap as a factor of production. Nor would a mass production technique recommend itself to a society that did not have a mass market. Indeed, the presence of slave labor seems generally to inhibit the use of machinery and the presence of expensive labor to accelerate it.[11]

These reflections on the social forces bearing on technical progress tempt us to throw aside the whole notion of technological determinism as false or misleading.[12] Yet, to relegate technology from an undeserved position of *primum mobile* in history to that of a mediating factor, both acted upon by and acting on the body of society, is not to write off its influence but only to specify its mode of operation with greater precision. Similarly, to admit we understand very little of the cultural factors that give rise to technology does not depreciate its role but focuses our attention on that period of history when technology is clearly a major historic force, namely Western society since 1700.

What is the mediating role played by technology within modern Western society? When we ask this much more modest question, the interaction of society and technology begins to clarify itself for us:

1 The rise of capitalism provided a major stimulus for the development of a technology of production

Not until the emergence of a market system organized around the principle of private property did there also emerge an institution capable of systematically guiding the inventive and innovative abilities of society to the problem of facilitating production. Hence the environment of the eighteenth and nineteenth centuries provided both a novel and an extremely effective encouragement for the development of an *industrial* technology. In addition, the slowly opening political and social framework of late mercantilist society gave rise to social aspirations for which the new technology offered the best chance of realization. It was not only the steam-mill that gave us the industrial capitalist but the rising inventor-manufacturer who gave us the steam-mill.

2 The expansion of technology within the market system took on a new 'automatic' aspect

Under the burgeoning market system not alone the initiation of technical improvement but its subsequent adoption and repercussion through the economy was largely governed by market considerations. As a result, both the rise and the proliferation of technology assumed the attributes of an impersonal diffuse 'force' bearing on social and economic life. This was all the more pronounced because the political control needed to buffer its disruptive consequences was seriously inhibited by the prevailing laissez-faire ideology.

3 **The rise of science gave a new impetus to technology**
The period of early capitalism roughly coincided with and provided a congenial setting for the development of an independent source of technological encouragement – the rise of the self-conscious activity of science. The steady expansion of scientific research, dedicated to the exploration of nature's secrets and to their harnessing for social use, provided an increasingly important stimulus for technological advance from the middle of the nineteenth century. Indeed, as the twentieth century has progressed, science has become a major historical force in its own right and is now the indispensable precondition for an effective technology.

It is for these reasons that technology takes on a special significance in the context of capitalism – or, for that matter, of a socialism based on maximizing production or minimizing costs. For in these societies, both the continuous appearance of technical advance and its diffusion throughout the society assume the attributes of autonomous process, 'mysteriously' generated by society and thrust upon its members in a manner as indifferent as it is imperious. This is why, I think, the problem of technological determinism – of how machines make history – comes to us with such insistence despite the ease with which we can disprove its more extreme contentions.

Technological determinism is thus peculiarly a problem of a certain historic epoch – specifically that of high capitalism and low socialism – *in which the forces of technical change have been unleased, but when the agencies for the control or guidance of technology are still rudimentary.*

The point has relevance for the future. The surrender of society to the free play of market forces is now on the wane, but its subservience to the impetus of the scientific ethos is on the rise. The prospect before us is assuredly that of an undiminished and very likely accelerated pace of technical change. From what we can foretell about the direction of this technological advance and the structural alterations it implies, the pressures in the future will be toward a society marked by a much greater degree of organization and deliberate control. What other political, social, and existential changes the age of the computer will also bring we do not know. What seems certain, however, is that the problem of technological determinism – that is, of the impact of machines on history – will remain germane until there is forged a degree of public control over technology far greater than anything that now exists.

Notes

1 See Robert K. Merton, 'Singletons and Multiples in Scientific Discovery: A Chapter in the Sociology of Science,' *Proceedings* of the American Philosophical Society, CV (October, 1961), 470–86.
2 See John Jewkes, David Sawers, and Richard Stillerman, *The Sources of Invention* (New York, 1960 [paperback edition]), p. 227, for a skeptical view.
3 'One can count 21 basically different means of flying, at least eight basic methods of geophysical prospecting; four ways to make uranium explosive . . . 20 or 30 ways to control birth . . . If each of these separate inventions were autonomous, i.e., without cause, how could one account for their arriving in these functional groups?' S.C. Gilfillan, 'Social Implications of Technological Advance,' *Current Sociology*, I (1952), 197. See also Jacob Schmookler, 'Economic Sources of Inventive Activity,' *Journal of Economic History* (March 1962), pp. 1–20; and Richard Nelson, 'The Economics of Invention:

A Survey of the Literature,' *Journal of Business*, XXXII (April 1959), 101–19.
4 Jewkes *et al.* (see n. 2) present a catalogue of chastening mistakes (p. 230 f.). On the other hand, for a sober predictive effort, see Francis Bello, 'The 1960s: A Forecast of Technology,' *Fortune*, LIX (January 1959), 74–78; and Daniel Bell, 'The Study of the Future,' *Public Interest*, I (Fall 1965), 119–30. Modern attempts at prediction project likely avenues of scientific advance or technological function rather than the feasibility of specific machines.
5 To be sure, the inquiry now regresses one step and forces us to ask whether there are inherent stages for the expansion of knowledge, at least insofar as it applies to nature. This is a very uncertain question. But having already risked so much, I will hazard the suggestion that the roughly parallel sequential development of scientific understanding in those few cultures that have cultivated it (mainly classical Greece, China, the high Arabian culture, and the West since the Renaissance) makes such a hypothesis possible, provided that one looks to broad outlines and not to inner detail.
6 The phrase is Richard LaPiere's in *Social Change* (New York, 1965), p. 263 f.
7 Karl Marx and Friedrich Engels, *The German Ideology* (London, 1942), p. 18.
8 Gilfillan (see n. 3), p. 202.
9 An interesting attempt to find a line of social causation is found in E. Hagen, *The Theory of Social Change* (Homewood, Ill., 1962).
10 See K.R. Gilbert, 'Machine-Tools,' in Charles Singer, E.J. Holmyard, A.R. Hall, and Trevor I. Williams (eds.), *A History of Technology* (Oxford, 1958), IV, chap. xiv.
11 See LaPiere (see n. 6), p. 294; also H. J. Habbakuk, *British and American Technology in the 19th Century* (Cambridge, 1962), *passim*.
12 As, for example, in A. Hansen, 'The Technological Determination of History,' *Quarterly Journal of Economics* (1921), pp. 76–83.

6.2 Jürgen Habermas, *Technology and Science as 'Ideology'*, 1971

By means of the concept of 'rationalization' [Max] Weber attempted to grasp the repercussions of scientific-technical progress on the institutional framework of societies engaged in 'modernization.' He shared this interest with the classical sociological tradition in general, whose pairs of polar concepts all revolve about the same problem: how to construct a conceptual model of the institutional change brought about by the extension of subsystems of purposive-rational action. [. . .]

In order to reformulate what Weber called 'rationalization,' I should like to propose another categorial framework. I shall take as my starting point the fundamental distinction between *work* and *interaction*.[1]

By 'work' or *purposive-rational action* I understand either instrumental action or rational choice or their conjunction. Instrumental action is governed by *technical rules* based on empirical knowledge. In every case they imply conditional predictions about observable events, physical or social. These predictions can prove correct or incorrect. The conduct of rational choice is governed by *strategies* based on analytic knowledge. They imply deductions from preference rules (value systems) and decision procedures; these propositions are either correctly or incorrectly deduced. Purposive-rational action realizes defined goals under given conditions. But while

Source: Jürgen Habermas, *Towards a Rational Society: Student Protest, Science and Politics*, trans. Jeremy J. Shapiro, London: Heinemann, 1971, pp. 81–122.

instrumental action organizes means that are appropriate or inappropriate according to criteria of an effective control of reality, strategic action depends only on the correct evaluation of possible alternative choices, which results from calculation supplemented by values and maxims.

By 'interaction,' on the other hand, I understand *communicative action*, symbolic interaction. It is governed by binding *consensual norms*, which define reciprocal expectations about behaviour and which must be understood and recognized by at least two acting subjects. Social norms are enforced through sanctions. Their meaning is objectified in ordinary language communication. While the validity of technical rules and strategies depends on that of empirically true or analytically correct propositions, the validity of social norms is grounded only in the intersubjectivity of the mutual understanding of intentions and secured by the general recognition of obligations. Violation of a rule has a different consequence according to type. *Incompetent* behaviour, which violates valid technical rules or strategies, is condemned per se to failure through lack of success; the 'punishment' is built, so to speak, into its rebuff by reality. *Deviant* behaviour, which violates consensual norms, provokes sanctions that are connected with the rules only externally, that is by convention. Learned rules of purposive-rational action supply us with *skills*, internalized norms with *personality structures*. Skills put us in a position to solve problems; motivations allow us to follow norms. [. . .]

In terms of the two types of action we can distinguish between social systems according to whether purposive-rational action or interaction predominates. The institutional framework of a society consists of norms that guide symbolic interaction. But there are subsystems such as (to keep to Weber's examples) the economic system or the state apparatus, in which primarily sets of purposive-rational action are institutionalized. These contrast with subsystems such as family and kinship structures, which, although linked to a number of tasks and skills, are primarily based on moral rules of interaction. So I shall distinguish generally at the analytic level between (1) the *institutional framework* of a society or the sociocultural life-world and (2) the *subsystems of purposive-rational action* that are 'embedded' in it. Insofar as actions are determined by the institutional framework they are both guided and enforced by norms. Insofar as they are determined by subsystems of purposive-rational action, they conform to patterns of instrumental or strategic action. Of course, only institutionalization can guarantee that such action will in fact follow definite technical rules and expected strategies with adequate probability.

With the help of these distinctions we can reformulate Weber's concept of 'rationalization.'

The term 'traditional society' has come to denote all social systems that generally meet the criteria of civilizations. The latter represent a specific stage in the evolution of the human species. They differ in several traits from more primitive social forms: (1) A centralized ruling power (state organization of political power in contrast to tribal organization); (2) The division of society into socioeconomic classes (distribution to individuals of social obligations and rewards according to class membership and not according to kinship status); (3) The prevalence of a central worldview (myth, complex religion) to the end of legitimating political power (thus converting power into authority). Civilizations are established on the basis of a relatively developed technology and of division of labour in the social process of production, which makes possible a surplus product, i.e. a quantity of goods exceeding that needed for the satisfaction of immediate and elementary needs. They owe their existence to the solution of the problem that first arises with the production of a surplus product, namely, how to distribute wealth and labor both unequally and yet legitimately according to criteria other than those generated by a kinship system.[2]

In our context it is relevant that despite considerable differences in their level of development, civilizations, based on an economy dependent on agriculture and craft production, have tolerated technical innovation and organizational improvement only within definite limits. One indicator of the traditional limits to the development of the forces of production is that until about three hundred years ago no major social system had produced more than the equivalent of a maximum of two hundred dollars per capita per annum. The stable pattern of a precapitalist mode of production, preindustrial technology, and premodern science makes possible a typical relation of the institutional framework to subsystems of purposive-rational action. For despite considerable progress, the subsystems, developing out of the system of social labour and its stock of accumulated technically exploitable knowledge, never reached that measure of extension after which their 'rationality' would have become an open threat to the authority of the cultural traditions that legitimate political power. The expression 'traditional society' refers to the circumstance that the institutional framework is grounded in the unquestionable underpinning of legitimation constituted by mythical, religious or metaphysical interpretations of reality – cosmic as well as social – as a whole. 'Traditional' societies exist as long as the development of subsystems of purposive-rational action keep within the limits of the legitimating efficacy of cultural traditions.[3] This is the basis for the 'superiority' of the institutional framework, which does not preclude structural changes adapted to a potential surplus generated in the economic system but does preclude critically challenging the traditional form of legitimation. This immunity is a meaningful criterion for the delimitation of traditional societies from those which have crossed the threshold to modernization.

The 'superiority criterion,' consequently, is applicable to all forms of class society organized as a state in which principles of universally valid rationality (whether of technical or strategic means-end relations) have not explicitly and successfully called into question the cultural validity of intersubjectively shared traditions, which function as legitimations of the political system. It is only since the capitalist mode of production has equipped the economic system with a self-propelling mechanism that ensures long-term continuous growth (despite crises) in the productivity of labour that the introduction of new technologies and strategies, i.e. innovation as such, has been institutionalized. As Marx and Schumpeter have proposed in their respective theories, the capitalist mode of production can be comprehended as a mechanism that guarantees the *permanent* expansion of subsystems of purposive-rational action and thereby overturns the traditionalist 'superiority' of the institutional framework to the forces of production. Capitalism is the first mode of production in world history to institutionalize self-sustaining economic growth. It has generated an industrial system that could be freed from the institutional framework of capitalism and connected to mechanisms other than that of the utilization of capital in private form.

What characterizes the passage from traditional society to society commencing the process of modernization is a level of development of the productive forces that makes permanent the extension of subsystems of purposive-rational action and thereby calls into question the traditional form of the legitimation of power. The older mythic, religious, and metaphysical worldviews obey the logic of interaction contexts. They answer the central questions of men's collective existence and of individual life history. Their themes are justice and freedom, violence and oppression, happiness and gratification, poverty, illness, and death. Their categories are victory and defeat, love and hate, salvation and damnation. Their logic accords with the grammar of systematically distorted communication and with the fateful causality of dissociated symbols and suppressed motives.[4] The rationality of language games, associated with communicative action, is confronted at the threshold of the modern period with the rationality of means-ends relations, associated with instrumental and strategic action. As soon as this confrontation can arise, the end of traditional society is in sight: the traditional form of legitimation breaks down.

Capitalism is defined by a mode of production that not only poses this problem but also solves it. It provides a legitimation of domination which is no longer called down from the lofty heights of cultural tradition but instead summoned up from the base of social labour. The institution of the market, in which private property owners exchange commodities – including the market on which propertyless private individuals exchange their labour power as their only commodity – promises that exchange relations will be and are just owing to equivalence. Even this bourgeois ideology of justice, by adopting the category of reciprocity, still employs a relation of communicative action as the basis of legitimation. But the principle of reciprocity is now the organizing principle of the sphere of production and reproduction itself. Thus on the base of a market economy, political domination can be legitimated henceforth 'from below' rather than 'from above' (through invocation of cultural tradition).

If we suppose that the division of society into socioeconomic classes derives from the differential distribution among social groups of the relevant means of production, and that this distribution itself is based on the institutionalization of relations of social force, then we may assume that in all civilizations this institutional framework has been identical with the system of political domination: traditional authority was political authority. Only with the emergence of the capitalist mode of production can the legitimation of the institutional framework be linked immediately with the system of social labour. Only then can the property order change from a *political relation* to a *production relation*, because it legitimates itself through the rationality of the market, the ideology of exchange society, and no longer through a legitimate power structure. It is now the political system which is justified in terms of the legitimate relations of production.

The superiority of the capitalist mode of production to its predecessors has these two roots: the establishment of an economic mechanism that renders permanent the expansion of subsystems of purposive-rational action, and the creation of an economic legitimation by means of which the political system can be adapted to the new requisites of rationality brought about by these developing subsystems. It is this process of adaptation that Weber comprehends as 'rationalization.' Within it we can distinguish between two tendencies: rationalization 'from below' and rationalization 'from above.'

A permanent pressure for adaptation arises from below as soon as the new mode of production becomes fully operative through the institutionalization of a domestic market for goods and labour power and of the capitalist enterprise. In the system of social labour this institutionalization ensures cumulative progress in the forces of production and an ensuing horizontal extension of subsystems of purposive-rational action – at the cost of economic crises, to be sure. In this way traditional structures are increasingly subordinated to conditions of instrumental or strategic rationality: the organization of labour and of trade, the network of transportation, information, and communication, the institutions of private law, and, starting with financial administration, the state bureaucracy. Thus arises the substructure of a society under the compulsion of modernization. The latter eventually widens to take in all areas of life: the army, the school system, health services, and even the family. Whether in city or country, it induces an urbanization of the *form* of life. That is, it generates subcultures that train the individual to be able to 'switch over' at any moment from an interaction context to purposive-rational action.

This pressure for rationalization coming from below is met by a compulsion to rationalize coming from above. For, measured against the new standards of purposive rationality, the power-legitimating and action-orienting traditions – especially mythological interpretations and religious worldviews – lose their cogency. On this level of generalization, what Weber termed 'secularization' has two aspects. First, traditional worldviews and objectivations lose their power and validity *as* myth, *as* public religion, *as* customary ritual, *as* justifying metaphysics, *as* unquestionable tradition. Instead, they are reshaped into subjective belief systems and ethics which ensure the private cogency of modern value-orientations (the 'Protestant ethic'). Second, they are transformed into constructions that do both at once: criticize tradition and reorganize the released material of tradition according to the principles of formal law and the exchange of equivalents (rationalist natural law). Having become fragile, existing legitimations are replaced by new ones. The latter emerge from the critique of the dogmatism of traditional interpretations of the world and claim a scientific character. Yet they retain legitimating functions, thereby keeping actual power relations inaccessible to analysis and to public consciousness. It is in this way that ideologies in the restricted sense first came into being. They replace traditional legitimations of power by appearing in the mantle of modern science and by deriving their justification from the critique of ideology. Ideologies are coeval with the critique of ideology. In this sense there can be no prebourgeois 'ideologies.'

In this connection modern science assumes a singular function. In distinction from the philosophical sciences of the older sort, the empirical sciences have developed since Galileo's time within a methodological frame of reference that reflects the transcendental viewpoint of possible technical control. Hence the modern sciences produce knowledge which through its *form* (and not through the subjective intention of scientists) is technically exploitable knowledge, although the possible applications generally are realized afterwards. Science and technology were not interdependent until late into the nineteenth century. Until then modern science did not contribute to the acceleration of technical development nor, consequently, to the pressure toward rationalization from below. Rather, its contribution to the modernization process was indirect. Modern physics gave rise to a philosophical approach that interpreted nature and society according to a model borrowed from the natural sciences and induced, so to speak, the mechanistic worldview of the seventeenth century. The reconstruction of classical natural law was carried out in this framework. This modern natural law was the basis of the bourgeois revolutions of the seventeenth, eighteenth, and nineteenth centuries, through which the old legitimations of the power structure were finally destroyed.[5]

By the middle of the nineteenth century the capitalist mode of production had developed so fully in England and France that Marx was able to identify the locus of the institutional framework of society in the relations of production and at the same time criticize the legitimating basis constituted by the exchange of equivalents. [. . .]

Since the last quarter of the nineteenth century two developmental tendencies have become noticeable in the most advanced capitalist countries: an increase in state intervention in order to secure the system's stability, and a growing interdependence of research and technology, which has turned the sciences into the leading productive force. Both tendencies have destroyed the particular constellation of institutional framework and subsystems of purposive-rational action which characterized liberal capitalism. I believe that Marcuse's basic thesis, according to which technology and science today also take on the function of legitimating political power, is the key to analyzing the changed constellation.

The permanent regulation of the economic process by means of state intervention arose as a defense mechanism against the dysfunctional tendencies, which threaten the system, that capitalism generates when left to itself. Capitalism's actual development manifestly contradicted the capitalist idea of a bourgeois society, emancipated from domination, in which power is neutralized. The root ideology of just exchange, which Marx unmasked in theory, collapsed in practice. The form of capital utilization through private ownership could only be maintained by the governmental corrective of a social and economic policy that stabilized the business cycle. The institutional framework of society was repoliticized. It no longer coincides immediately with the relations of production, ie. with an order of private law that secured capitalist economic activity and the corresponding general guarantees of order provided by the bourgeois state. But this means a change in the relation of the economy to the political system: politics is no longer *only* a phenomenon of the superstructure. If society no longer 'autonomously' perpetuates itself through self-regulation as a sphere preceding and lying at the basis of the state – and its ability to do so was the really novel feature of the capitalist mode of production – then society and the state are no longer in the relationship that Marxian theory had defined as that of base and superstructure. Then, however, a critical theory of society can no longer be constructed in the exclusive form of a critique of political economy. A point of view that methodically isolates the economic laws of motion of society can claim to grasp the overall structure of social life in its essential categories only as long as politics depends on the economic base. It becomes inapplicable when the 'base' has to be comprehended as in itself a function of governmental activity and political conflicts. According to Marx, the critique of political economy was the theory of bourgeois society only as *critique of ideology*. If, however, the ideology of just exchange disintegrates, then the power structure can no longer be criticized *immediately* at the level of the relations of production.

With the collapse of this ideology, political power requires a new legitimation. Now since the power indirectly exercised over the exchange process is itself operating under political control and state regulation, legitimation can no longer be derived from the unpolitical order constituted by the relations of production. To this extent the requirement for direct legitimation, which exists in precapitalist societies, reappears. On the other hand, the resuscitation of immediate political domination (in the traditional form of legitimation on the basis of cosmological worldviews) has become impossible. For traditions have already been disempowered. Moreover, in industrially developed societies the results of bourgeois emancipation from immediate political domination (civil and political rights and the mechanism of general elections) can be fully ignored only in periods of reaction. Formally democratic government in systems of state-regulated capitalism is subject to a need for legitimation which cannot be met by a return to a prebourgeois form. Hence the ideology of free exchange is replaced by a substitute program. The latter is oriented not to the social results of the institution of the market but to those of government action designed to compensate for the dysfunctions of free exchange. This policy combines the element of the bourgeois ideology of achievement (which, however, displaces assignment of status according to the standard of individual achievement from the market to the school system) with a guaranteed minimum level of welfare, which offers secure employment and a stable income. This substitute program obliges the political system to maintain stabilizing conditions for an economy that guards against risks to growth and guarantees social security and the chance for individual upward mobility. What is needed to this end is latitude for manipulation by state interventions that, at the cost of limiting the institutions of private law, secure the private form of capital utilization *and bind the masses' loyalty to this form.*

Insofar as government action is directed toward the economic system's stability and growth, politics now takes on a peculiarly negative character. For it is oriented toward the elimination of dysfunctions and the avoidance of risks that threaten the system: not, in other words, toward the *realization of practical goals* but toward the *solution of technical problems*.
[. . .] [G]overnment activity is restricted to administratively soluble technical problems, so that practical questions evaporate, so to speak. *Practical substance is eliminated.*

Old-style politics were forced, merely through its traditional form of legitimation, to define itself in relation to practical goals: the 'good life' was interpreted in a context defined by interaction relations. The same still held for the ideology of bourgeois society. The substitute program prevailing today, in contrast, is aimed exclusively at the functioning of a manipulated system. It eliminates practical questions and therewith precludes discussion about the adoption of standards; the latter could emerge only from a democratic decision-making process. The solution of technical problems is not dependent on public discussion. Rather, public discussions could render problematic the framework within which the tasks of government action present themselves as technical ones. Therefore the new politics of state interventionism requires a depoliticization of the mass of the population. To the extent that practical questions are eliminated, the public realm also loses its political function. At the same time, the institutional framework of society is still distinct from the systems of purposive-rational action themselves. Its organization continues to be a problem of *practice* linked to communication, not one of *technology*, no matter how scientifically guided. Hence, the bracketing out of practice associated with the new kind of politics is not automatic. The substitute program, which legitimates power today, leaves unfilled a vital need for legitimation: how will the depoliticization of the masses be made plausible to them? Marcuse would be able to answer: by having technology and science *also* take on the role of an ideology.

Since the end of the nineteenth century the other developmental tendency characteristic of advanced capitalism has become increasingly momentous: the scientization of technology. The institutional pressure to augment the productivity of labour through the introduction of new technology has always existed under capitalism. But innovations depended on sporadic inventions, which, while economically motivated, were still fortuitous in character. This changed as technical development entered into a feedback relation with the progress of the modern sciences. With the advent of large-scale industrial research, science, technology, and industrial utilization were fused into a system. Since then, industrial research has been linked up with research under government contract, which primarily promotes scientific and technical progress in the military sector. From there information flows back into the sectors of civilian production. Thus technology and science become a leading productive force, rendering inoperative the conditions for Marx's labour theory of value. It is no longer meaningful to calculate the amount of capital investment in research and development on the basis of the value of unskilled (simple) labour power, when scientific-technical progress has become an independent source of surplus value, in relation to which the only source of surplus value considered by Marx, namely the labour power of the immediate producers, plays an ever smaller role.

As long as the productive forces were visibly linked to the rational decisions and instrumental action of men engaged in social production, they could be understood as the potential for a growing power of technical control and not be confused with the institutional framework in which they are embedded. However, with the institutionalization of scientific-technical progress, the potential of the productive forces has assumed a form owing to which men lose consciousness of the dualism of work and interaction.

It is true that social interests still determine the direction, functions, and pace of technical progress. But these interests define the social system so much as a whole that they coincide with the interest in maintaining the system. *As such* the private form of capital utilization and a distribution mechanism for social rewards that guarantees the loyalty of the masses are removed from discussion. The quasi-autonomous progress of science and technology then appears as an independent variable on which the most important single system variable, namely economic growth, depends. Thus arises a perspective in which the development of the social system *seems* to be determined by the logic of scientific-technical progress. The immanent law of this progress seems to produce objective exigencies, which must be obeyed by any politics oriented toward functional needs. But when this semblance has taken root effectively, then propaganda can refer to the role of technology and science in order to explain and legitimate why in modern societies the process of democratic decision-making about practical problems loses its function and 'must' be replaced by plebiscitary decisions about alternative sets of leaders of administrative personnel. This technocracy thesis has been worked out in several versions on the intellectual level.[6] What seems to me more important is that it can also become a background ideology that penetrates into the consciousness of the depoliticized mass of the population, where it can take on legitimating power.[7] It is a singular achievement of this ideology to detach society's self-understanding from the frame of reference of communicative action and from the concepts of symbolic interaction and replace it with a scientific model. Accordingly the culturally defined self-understanding of a social life-world is replaced by the self-reification of men under categories of purposive-rational action and adaptive behaviour. [. . .]

[T]he leading productive force – controlled scientific-technical progress itself – has now become the basis of legitimation. Yet this new form of legitimation has cast off the old shape of *ideology*.

Technocratic consciousness is, on the one hand, 'less ideological' than all previous ideologies. For it does not have the opaque force of a delusion that only transfigures the implementation of interests. On the other hand today's dominant, rather glassy background ideology, which makes a fetish of science, is more irresistible and farther-reaching than ideologies of the old type. For with the veiling of practical problems it not only justifies a *particular class's* interest in domination and represses *another class's* partial need for emancipation, but affects the human race's emancipatory interest as such.

Technocratic consciousness is not a rationalized, wish-fulfilling fantasy, not an 'illusion' in Freud's sense, in which a system of interaction is either represented or interpreted and grounded. Even bourgeois ideologies could be traced back to a basic pattern of just interactions, free of domination and mutually satisfactory. It was these ideologies which met the criteria of wish-fulfillment and substitute gratification; the communication on which they were based was so limited by repressions that the relation of force once institutionalized as the capital-labour relation could not even be called by name. But the technocratic consciousness is not based in the same way on the causality of dissociated symbols and unconscious motives, which generates both false consciousness and the power of reflection to which the critique of ideology is indebted. It is less vulnerable to reflection, because it is no longer *only* ideology. For it does not, in the manner of ideology, express a projection of the 'good life' (which even if not identifiable with a bad reality, can at least be brought into virtually satisfactory accord with it). Of course the new ideology, like the old, serves to impede making the foundations of society the object of thought and reflection. Previously, social force lay at the basis of the relation between capitalist and wage-labourers. Today the basis is provided by structural conditions which predefine the tasks of system maintenance: the private form of capital utilization and a political form of distributing social rewards that guarantees mass loyalty. However, the old and new ideology differ in two ways.

First, the capital-labour relation today, because of its linkage to a loyalty-ensuring political distribution mechanism, no longer engenders uncorrected exploitation and oppression. Technocratic consciousness, therefore, cannot rest in the same way on collective repression as did earlier ideologies. Second, mass loyalty today is created only with the aid of rewards for *privatized needs*. The achievements in virtue of which the system justifies itself may not in principle be interpreted politically. The acceptable interpretation is immediately in terms of allocations of money and leisure time (neutral with regard to their use), and mediately in terms of the technocratic justification of the occlusion of practical questions. Hence the new ideology is distinguished from its predecessor in that it severs the criteria for justifying the organization of social life from any normative regulation of interaction, thus depoliticizing them. It anchors them instead in functions of a putative system of purposive-rational action.

Technocratic consciousness reflects not the sundering of an ethical situation but the repression of 'ethics' as such as a category of life. The common, positivist way of thinking renders inert the frame of reference of interaction in ordinary language, in which domination and ideology both arise under conditions of distorted communication and can be reflectively detected and broken down. The depoliticization of the mass of the population, which is legitimated through technocratic consciousness, is at the same time men's self-objectification in categories equally of both purposive-rational action and adaptive behavior. The reified models of the sciences migrate into the sociocultural life-world and gain objective power over the latter's self-understanding. The ideological nucleus of this consciousness is *the elimination of the distinction between the practical and the technical.* It reflects, but does not objectively account for, the new constellation of a disempowered institutional framework and systems of purposive-rational action that have taken on a life of their own. [. . .]

Notes

1 On the context of these concepts in the history of philosophy, see my contribution to the *Festschrift* for Karl Löwith: 'Arbeit und Interaktion: Bemerkungen zu Hegels Jenenser Realphilosophie,' in *Natur und Geschichte, Karl Löwith zum 70. Geburtstag*, Hermann Braun and Manfred Riedel, eds. (Stuttgart, 1967). This essay is reprinted in *Technik und Wissenschaft als 'Ideologie'* (Frankfurt am Main, 1968) and in English in *Theory and Practice* [Beacon Press, 1973].

2 Gerhard E. Lenski, *Power and Privilege: A Theory of Social Stratification* (New York, 1966).

3 See Peter L. Berger, *The Sacred Canopy* (New York, 1967).

4 See my study *Erkenntnis und Interesse* (Frankfurt am Main, 1968); [published in English as *Knowledge and Human Interests*, Beacon Press, 1971].
5 See Jürgen Habermas, 'Naturrecht und Revolution,' in *Theorie und Praxis*.
6 See [for example] Jacques Ellul, *The Technological Society* (New York, 1967).
7 To my knowledge there are no empirical studies concerned specifically with the propagation of this background ideology. We are dependent on extrapolations from the findings of other investigations.

Index